SpringerBriefs in Water Science and Technology

More information about this series at http://www.springer.com/series/11214

Ganesh Keremane

Governance of Urban Wastewater Reuse for Agriculture

A Framework for Understanding and Action in Metropolitan Regions

 Springer

Ganesh Keremane
School of Law, UniSA Business School
University of South Australia
Adelaide, SA
Australia

ISSN 2194-7244 ISSN 2194-7252 (electronic)
SpringerBriefs in Water Science and Technology
ISBN 978-3-319-55055-8 ISBN 978-3-319-55056-5 (eBook)
DOI 10.1007/978-3-319-55056-5

Library of Congress Control Number: 2017933851

Printed on acid-free paper

This Springer imprint is published by Springer Nature
The registered company is Springer International Publishing AG
The registered company address is: Gewerbestrasse 11, 6330 Cham, Switzerland

Preface

Freshwater scarcity has engendered two immediate responses: different water allocation methods, and development and use of alternative sources of water. While water markets are seen as a means to achieve efficient allocation of the scare resources, urban wastewater reuse (for non-potable applications including agricultural irrigation) appears as a viable option to augment traditional water supplies. Additionally, with the 'fit-for-purpose' argument surfacing on the global water governance agenda, the search for a reliable alternative source of water has triggered governmental support for the development of water reclamation and reuse laws, policies, and projects in many countries. As such, water recycling or use of recycled water (for non-potable applications) has assumed a recognized and important role in the portfolio of urban water management strategies around the world. As the level of water recycling increases, the choice and implementation of alternative policy instruments, governance arrangements, and incentives to assist in the promotion and coordination of water recycling also assume increasing importance. Among other factors, decision support in policy design and implementation (institutions and governance) is a key to achieving water sustainability. Institutions and governance frameworks will need to provide for the rights of access, rights of ownership, rights to manage source and treated water, and the obligations of final use of recycling operations. The primary focus of this book is not on the technical aspects of designing and building infrastructure. Rather, it seeks to provide guidance to better understand the institutional and governance challenges of managing urban wastewater, particularly for reuse in agriculture.

This book is one of the main outputs of a Ph.D. project which has gathered and synthesized knowledge from Australia and India on governance paradigms and institutional arrangements for urban wastewater reuse in these countries, specifically in two metropolitan areas: Adelaide (Australia) and Hyderabad (India). Using three case studies representing different models of governance, this book analyses the role of different societal sectors—public, private, and the community in provision and use of wastewater for irrigation. This book is, therefore, not intended as technical manual for engineers or planners involved in designing or building water/wastewater infrastructure. Instead, it is designed to help users systematically examine the institutional and governance issues that influence the implementation of urban wastewater reuse projects.

Lastly, literature on wastewater reuse mostly comprises studies that have adopted a scientific and biophysical approach, and there is lack of institutional studies using a combination of social, quantitative, and qualitative methodologies. This impedes the formulation of recommendations that could enhance the benefits and ease the concerns of all groups involved with wastewater reuse. Furthermore, these studies can be carried out at different levels—macro-, meso-, and microlevels. The mesolevel includes the wastewater delivery or supply system, which is the largest element of the complex system, and the unit of analysis at the microlevel includes the beneficiaries/households and those local institutions that shape the wastewater use. Accordingly, this book adopts an institutional approach and focuses at both the meso- and microlevels of analysis, thereby contributing to the literature.

This book is organized into nine chapters. Chapter 1 sets the context and scope of the research study. Chapter 2 provides an account of urban wastewater reuse and its applications and discusses the challenges facing policy makers and water managers as they implement wastewater reuse projects while Chap. 3 focuses on the water governance regimes and wastewater reuse in Australia and India. Chapter 4 provides the theoretical background as various theories related to water governance are discussed and the interrelationship between these theories are examined to provide a framework for analysing the institutional frameworks and regulations governing the use of urban wastewater for agriculture. Chapter 5 describes the research methods and introduces the case study sites in Australia and India, and explains the criteria adopted to select the schemes, respondents, and key stake-holders. Chapters 6 through 8 discusses the results of the three case studies separately as they all have varying governance or organizational structures, and are examples of wastewater reuse that rely on cohesive local networks and involvement of all three societal sectors—public, private, and community. Chapter 9 presents the conclusions drawn from the analysis of the three case studies in Australia and India. It covers the theoretical and empirical conclusions, followed by the recommendations and policy options for wastewater reuse in agriculture.

Adelaide, Australia   Ganesh Keremane

Acknowledgements

The present book is an abridged version of my thesis which was accepted by the School of Commerce of the University of South Australia in partial fulfilment of the requirements for the degree of Ph.D. I would like to express my sincere appreciation to all those who have directly or indirectly contributed to this thesis.

I would like to thank my Principal Supervisor Prof. Jennifer McKay for her academic supervision, professional guidance, personal support, and for reviewing the thesis. I would also like to thank Prof. Mervyn Lewis for reviewing the thesis as the associate supervisor.

I am greatly indebted to the following organizations: the University of South Australia for awarding me the President's Scholarship to pursue my Ph.D.; the Cooperative Research Centre for Irrigation Futures (CRC IF) for providing me with financial assistance to conduct the field work at both of the study sites in South Australia; and the Australian Centre for International Agricultural Research (ACIAR) for funding my research in India.

I would like to thank the whole staff—academic and professional—of the Centre for Comparative Water Policies and Laws for providing the best conditions for research. I would also like to thank all the people—research participants, stakeholders, interpreters—who helped me carry out the three case studies efficiently.

I would like to thank Fritz Schmuhl, senior publishing editor, Springer, for accepting the publication of the book in the Springer Briefs series.

I would like to thank Geoffrey for proofreading an earlier version of the text.

Finally, and on a more personal level, I sincerely thank my late parents and family who have always encouraged and supported my choices.

Last, but by no means the least, I would like to thank my wife Shwetha for continuously and patiently supporting me in every phase of the work on this thesis, and our son Nihal, who joined us during this adventure. I dedicate this book to them.

Adelaide, Australia   Ganesh Keremane
December 2016

Contents

Chapter 1
Introduction

Water is essential to the well-being of human kind, vital for economic development, and a basic requirement for the healthy functioning of all the world's ecosystems. While there are sufficient freshwater resources to meet everyone's basic personal and domestic needs, the extent to which people have access to these resources for various uses is limited. Reasons for this include: lack of distribution networks, excessive extraction of groundwater resources, and risk from the contamination by the pollutants. While there are some signs of greater efficiency in water use, the current indications of water use and management point out that the situation on the ground is getting worse and not better. Water withdrawal statistics indicate that annual global water withdrawals have increased by more than six times and the largest proportion of this growth is in countries with developing or emerging economies and increasing standards of living (UNESCO 2003). In some places groundwater levels continue to fall and the options for increasing supplies have become expensive and are often environmentally damaging (Frederick 2001). Furthermore, rapid urbanisation and industrialisation has resulted in the squeeze on freshwater supplies for agricultural uses and this necessitates we look for reliable, alternative sources of supply. Consequently, the water crisis has engendered new directions for water governance and development and use of urban wastewater as an alternative source of supply.

1.1 Urban Wastewater—a Reliable Alternative Source of Water

Agriculture is the largest consumer of freshwater resources, currently accounting for about 70% water withdrawals globally and over 90% in the developing world (UNESCO 2003). With increasing population growth, urbanisation, and rapid industrial development, the availability of freshwater is likely to be one of the major

© The Author(s) 2017
G. Keremane, *Governance of Urban Wastewater Reuse for Agriculture*,
SpringerBriefs in Water Science and Technology,
DOI 10.1007/978-3-319-55056-5_1

limits to economic development in the decades to come. It is expected that water now used for agriculture will be diverted to the urban and industrial sectors (Serageldin 1995) demanding to find a 'new' and 'reliable' source of supply to augment freshwater supplies, thereby reducing the pressure on existing resources. One way of responding to this squeeze on freshwater supply, particularly in the agriculture sector is by reuse of (treated) urban wastewater[1] for irrigation.

Wastewater reuse for non-potable purposes particularly for irrigation is a centuries-old practice. But it has been little reported or documented because the norm is to treat wastewater before use (Ensink et al. 2002). It's only in the recent past that due to sever water scarcity challenges development of water reclamation and reuse projects have received much impetus. Accordingly, reuse of (treated) wastewater for irrigation has increased overtime and will continue to increase in future. For example, in Israel and Palestine treated sewage effluent will become the main source of water for irrigation, supplying 1000 million m3 (70%) out of the 1400 million m3 that will be used for irrigation by the year 2040 (Israel Irrigation Commission 1995, cited in Haruvy et al. 1999, p. 303). Nevertheless, in many developing countries wastewater (mostly untreated) is a highly important productive resource, and is a substantial and sometimes even primary source of cash income for thousands of small farmers and the landless (Scott et al. 2000, 2004). The reasons for this include: increasing water scarcity, lack of funds for treatment, and a clear willingness by farmers to use untreated wastewater (Ensink et al. 2002).

Among the different applications of wastewater, it is believed that agricultural irrigation is the best use of wastewater after treatment (Pescod 1992), and the presence of crop nutrients in wastewater benefits crop production (Ensink et al. 2002). However, due to the regular concerns raised over the potential health impacts of using untreated wastewater its usage cannot be encouraged. Neither can we impose absolute restrictions considering the amount of pressure on existing freshwater supplies. So, to address these issues, wastewater guidelines constituting a common vision and direction for wastewater management need to be developed, like the Hyderabad Declaration on Wastewater Use in Agriculture signed by representatives of international and national institutions on 14 November 2002 at a global workshop in Hyderabad, India. The key message of the Declaration is to safeguard and strengthen livelihoods and food security, mitigate health and environmental risks and conserve water resources by confronting the realities of wastewater use in agriculture through the adoption of appropriate policies and the commitment of financial resources for policy implementation.

Urban wastewater reuse experiments around the world have demonstrated the feasibility of water reuse on a large scale and its role in the sustainable management of the world's water (Anderson 2003, p. 2). For example, in Israel, due to scarce water resources and the deteriorating quality of waters, the situation demanded a national policy recommending reuse of all municipal wastewater (Brenner et al. 2000).

---

[1]A combination of domestic effluent, water from commercial establishment and institutions, industrial effluent and storm water and other urban runoff.

Similarly, many countries, such as Singapore, Namibia, Mexico, Vietnam, China, Japan, Australia, the USA, and some European countries, have seen replacement of freshwater by treated wastewater as an important conservation strategy contributing to agricultural production and have successfully implemented direct and indirect water reuse projects (Ensink et al. 2002; Po et al. 2004). However, there is apprehension in the world community about direct potable reuse due to uncertainties about water quality and negative public perceptions (Hurlimann and Mckay 2006).

Despite all the advantages this resource has to offer, developing a sustainable wastewater reuse scheme is an onerous task; mostly because wastewater management spans a wide range of institutions and stakeholders which requires coordination of both policies and regulation governing the resource. An effective institutional network and a favourable regulatory and policy regime for wastewater management are essential to improve the acceptability of the scheme and delivering high value to the community and the environment. All these points highlight the importance of effective water governance and the institutional framework.

1.2   New Directions for (Waste)Water Governance

Water governance is a significant aspect of international development policy making. The United Nations World Water Development Report-2 (UNESCO 2006) recognizes that water crisis is largely a crisis of governance, and outlines many of the leading obstacles to sound and sustainable water management. There is an increasing consensus on the need to improve water governance to achieve the Millennium Development Goals (Institute of Development Studies 2007). Current situation demands a change or shift in water governance which Gleick (2000) describes as 'the changing water paradigm'.

The concept of governance has been widely debated since the 1990s and there are various definitions of this concept and approaches to it. Governance according to Stoker (1998, p. 17) "is ultimately concerned with creating the conditions for ordered rule and collective action". Kooiman (2003) provides a relatively broad definition of governance and describes it as,

> the totality of interactions, in which public as well as private actors participate, aimed at solving societal problems or creating societal opportunities; attending to the institutions as contexts for these governing interactions; and establishing a normative foundation for all those activities (p. 4).

Extending these views to water governance, Rogers and Hall (2003) define water governance as:

> the range of political, social, economic and administrative systems that are in place to develop and manage water resources, and the delivery of water services, at different levels of society (p. 16).

The authors further argue that water governance encompasses a large spectrum of aspects related to water and according to them:

the notion of governance for water includes the ability to design public policies and institutional frameworks that are socially acceptable and mobilise social resources in support of them. Water policy and the process for its formulation must have sustainable development of water resources as its goal, and to make it effective the key actors must be involved in the process (p. 16).

Drawing from the definitions cited above, water governance can be understood as a framework of political, social, economic, and legal structures within which societies choose and accept to manage their water-related affairs. It includes governments, the market forces that help to allocate resources, and any other mechanism that regulates human interactions. In simpler terms, water governance is the ongoing process of extracting, distributing, and using water created by the actors' purposeful actions within the present institutions or the rules-in-use.

1.2.1 The Shift in Water Governance Paradigm

Many countries have had profound policy changes in recent years, referred to by scholars as shifts in the policy paradigm (Menahem 1998). In public administration, the New Public Management (Larbi 1999) is one of the much-discussed changed paradigms. In the water sector, there has been a similar change in paradigm (Gleick 2000).

Water comes in many forms, with economic, social, religious, cultural and environmental values attached that are often interdependent; it must be shared between different uses and different users. So, governing water wisely is vitally important for sustainable water resource development; and the focus of governance in this sector needs to be shifted from 'water resource development' to 'water resource management'.

Traditionally, water management responsibilities have been vested with State or public agencies, with the assumption that public agencies possess all the necessary resources, expertise, and authority to manage this resource. However, policy makers and water planners now recognize and agree that public management has often failed to follow the basic principles of effective governance (UNESCO 2003). While acknowledging the failure of public agencies to manage the resources in question, it is also accepted that user groups or communities can manage the resource more effectively than can the public agencies. Following this, there has been a noticeable policy shift, in the form of partial or complete transfer of management responsibilities from the public agencies to user groups (Ostrom 1999; Tang 1992; Meinzen-Dick and Sullins 1994; Rasmussen and Meinzen-Dick 1995; Meinzen-Dick and Knox 1999; Agarwal and Ostrom 1999; Ostrom 2000; Holm-Muller and Zavgorodnyaya 2003). Consequently, water users' associations have played important roles in facilitating effective water management, and their roles generally fit into two broad categories—(a) mobilizing and organizing the community of water users, officials, and professionals in support of management initiatives, and (b) providing communication and dissemination of information and technical assistance that is beneficial for water management (Blomquist 1994).

On the other hand, inefficiency, corruption and lack of funds within the public utilities for extending access to services within the water and sanitation sector, have prompted increased private sector participation in addressing these problems. Although, private sector participation was strongly promoted on the water and sanitation policy agenda during the 1990s (Budds and McGranahan 2003), its prominence in the water sector remains limited. Private sector participation generally refers to contractual agreements between a public sector (government) and private agencies that can range from large water companies (usually multinational) to small-scale informal operators or civil societies. Likewise, the forms or models of private sector involvement vary according to the allocation of responsibilities and so experts have various opinions about water privatization. Under such circumstances the Integrated Water Resource Management (IWRM) approach is a comprehensive approach to the development and management of water. Allan (2001) while discussing the paradigms that have determined the way water resources have been perceived and managed in the twentieth century argues that IWRM requires a holistic approach and an unprecedented level of political cooperation.

On the other hand, Gleick (2000) emphasizes that new paradigm for water planning places a high value on maintaining the integrity of water resources, and the flora, fauna and human societies that are built around them. He further indicates that along with increasing the water allocation efficiency, development and use of non-traditional sources of supply (reclaimed/recycled water) will play an increasing role in the water management agenda. Accordingly, while water markets are believed to achieve efficient allocation of scarce resources, reuse of urban (treated) wastewater is being considered as a viable method for augmenting traditional water supplies. This book focuses on the latter option which is source substitution and discusses the use of treated urban wastewater in agriculture from an institutional analysis perspective.

1.3 Water Scarcity Crisis—Is Source Substitution the Answer?

One of the latest crises of modernity is water scarcity and 'source substitution' is increasingly being considered as a viable solution to our water supply challenges. While it is not the only solution to address the problems of water scarcity it certainly has become an integral part of water management policy in many water-scarce countries. This is largely because wastewater from point sources, such as sewage treatment plants and industries, provides an excellent source of reusable water and is usually available on a reliable basis, has a known quality, and can be accessed at a single point (Davis and Hirji 2003). Furthermore, urban wastewater use reduces the amount of waste discharged into watercourses and hence improves the environment. It also conserves water resources by lowering the demand for freshwater withdrawal (Khouri et al. 1994).

Source substitution allows the higher quality water to be used for domestic supply, and provides a suitable alternative for less critical uses (Hespanhol 1997). Accordingly, water reuse on a large scale is now an option in many areas and the fitness-for-purpose argument is on the global agenda of water governance. Furthermore, the search for reliable alternative sources of water has triggered governmental support for the development of water reclamation and reuse laws and policies, subsequently leading to practical projects in many countries. Concepts such as water reclamation, recycling and reuse have become the key components of water and wastewater management policies in many water scare nations. The many drivers of these concepts include the availability of alternative users and communities willing to use the water, prevention of environmental degradation, water conservation and economic advantages.

However, the idea has not been always well accepted by the community and the formal water supply institutions in developed countries have been hesitant and have involved the private sector to provide the actual services. Though it is believed that wastewater reuse can augment freshwater supplies, and help communities accrue substantial benefits, the development of sustainable water reuse schemes often encounter technical, financial, commercial, regulatory, policy, social and institutional impediments (Davis and Hirji 2003; Thiyagarajah 2005; Dimitriadis 2005). While a lot of work has been done on the technical aspect of wastewater reuse, the social and policy side has not received the required attention. Colebatch (2006) while examining the context of recycling as an institutional challenge raises a series of questions:

...how recycling can find a place in an organizational world built around an industrial paradigm of supply and disposal. Is the existing organisation to change its character, or is recycling to be added on? Is recycling to be accomplished centrally or does it need to be done at household or neighbourhood level, in which case, what organisational base is needed? What place would other stakeholders, such as health authorities or local government, have in these arrangements? How would the users be integrated into the structure? (pp. 24–25).

These questions clearly indicate that there is a need to direct our thinking towards the institutional dimension of water reuse/recycling which is less evident in the literature. Accordingly, this book seeks to provide guidance to better understand the institutional and governance challenges of managing urban wastewater reuse, particularly in agriculture. This is achieved by comparing case studies of urban wastewater reuse in Australia and India. The focus is largely on the processes of governance and institution formation for urban wastewater reuse in these cases, and not on the technical or financial aspects of designing and building infrastructure.

References

Agarwal A, Ostrom E (1999) Collective action, property rights, and devolution of forest and protected area management. Paper presented at Workshop on collective action, property rights and devolution of natural resources management, Puerto Azul, The Philippines, 21–25 June 1999

Allan T (2001) Millennial water management paradigms: making Integrated Water Resources Management (IWRM) work. Occasional Paper, Published by African Water Issues Research Unit, Pretoria, South Africa

Anderson J (2003) The environmental benefits of water recycling and reuse. Water Sci Technol: Water Supply 3(4):1–10

Blomquist W (1994) The local groundwater economy in Los Angeles County, California. Presented at the Workshop in political theory and policy analysis, Indiana University, Bloomington, Indiana, USA, 16–18 June 1994

Brenner A, Shandalov S, Messalem R, Yakirevich A, Oron G, Rebhun M (2000) Wastewater reclamation for agricultural reuse in Israel: trends and experimental results. Water Air Soil Pollut 123:167–182

Budds J, McGranahan G (2003) Are the debates on water privatization missing the point? Experiences from Africa, Asia and Latin America. Environ Urbanization 15:87–113

Colebatch HK (2006) Governing the use of water: the institutional context. Desalination 187:17–27

Davis R, Hirji R (eds) (2003) Water resources and environment–wastewater reuse. Technical Note, F.3. The World Bank, Washington, DC

Dimitriadis S (2005) Issues encountered in advancing Australia's water recycling schemes. (Research Brief, No. 2). Department of Parliamentary Services, Commonwealth of Australia, 2005–2006

Ensink JHJ, van der Hoek W, Matsuno Y, Munir S, Aslam RM (2002) Use of untreated wastewater in peri-urban agriculture in Pakistan: risks and opportunities, (Research Report-67). International Water Management Institute, Colombo, Sri Lanka

Frederick KD (2001) Water marketing: obstacles and opportunities. Forum Appl Res Public Policy 16(1):54–62

Gleick P (2000) The changing water paradigm – a look into twenty-first century water resources development. Water Int 25(1):127–138

Haruvy N, Offer R, Hadas A, Ravina I (1999) Wastewater irrigation-economic concerns regarding beneficiary and hazardous effects of nutrients. Water Resour Manage 13(5):303–314

Hespanhol I (1997) Wastewater as a resource. In: Helmer R, Hespanhol I (eds) Water pollution control: a guide to use of water quality management principles, published by E. & F. Spon on behalf of WHO/UNEP

Holm-Muller K, Zavgorodnyaya D (2003) Conflict-resolution mechanisms in Uzbek Water Users' Associations: one of the important institutional criteria. Paper presented at the 20th ICID European Regional Conference, Montpellier, 14–19 Sept 2003

Hurlimann A, McKay JM (2006) What attributes of recycled water make it fit for residential purposes? The Mawson Lakes experience. Desalination 187:167–177

Institute of Development Studies (2007) New directions for water governance. In: Woods Tim (ed) id21 insights 67, June 2007. University of Sussex, UK, pp 1–6

Khouri N, Kalbermatten JH, Bartone CR (1994) The reuse of wastewater in agriculture: a guide for planners. (Water and Sanitation Report No. 6). UNDP-World Bank Water and Sanitation Program, the World Bank, Washington, DC

Kooiman J (2003) Governing as governance. SAGE Publications, UK-USA-India

Larbi GA (1999) The new public management approach and crisis states. (Discussion Paper No. 112). UNRISD, Geneva, Switzerland, Sept 1999

Meinzen-Dick R, Knox A (1999) Collective action, property rights, and devolution of natural resource management: a conceptual framework. Paper presented at the Workshop on collective action, property rights and devolution of natural resources management, Puerto Azul, The Philippines, 21–25 June 1999

Meinzen-Dick R, Sullins M (1994) Water markets in Pakistan: participation and productivity. (EPTD Working Paper No. 4). IFPRI, Washington, DC

Menahem G (1998) Policy paradigms, policy networks and water policy in Israel. J Public Policy 18(3):283–310

Ostrom E (2000) Reformulating the commons. Swiss Polit Sci Rev 6(1):29–52

Ostrom E (1999) Coping with the tragedies of the commons. Annu Rev Polit Sci 2:493–535

Pescod MB (1992) Wastewater treatment and use in agriculture. Irrigation and drainage, (Paper No. 47). FAO, p 118

Po M, Juliane K, Nancarrow BE (2004) Literature review of factors influencing public perceptions of water reuse. Australian Water Conservation and Reuse Research Program, CSIRO

Rasmussen LN, Meinzen-Dick R (1995) Local organizations for natural resource management: lessons from theoretical and empirical literature. (ETPD Discussion Paper No. 11). IFPRI, Washington, DC

Rogers P, Hall AW (2003) Effective water governance. TEC Background Paper (No. 7). Global Water Partnership-TEC, Sweden

Scott CA, Zarazúa JA, Levine G (2000) Urban-wastewater reuse for crop production in the water-short Guanajuato river basin, Mexico. (Research Report No. 41). International Water Management Institute, Colombo, Sri Lanka

Scott C, Faruqui NI, Raschid L (eds) (2004) Wastewater use in irrigated agriculture–confronting the livelihoods and environmental realities. CABI/IWMI/IDRC, 2004

Serageldin I (1995) Toward sustainable management of water resources. World Bank, Washington, DC

Stoker G (1998) Governance as theory: five propositions. Int Soc Sci J 50(155):17–28

Tang SY (1992) Institutions and collective action: self-governance in irrigation. ICS Press, San Francisco, CA

Thiyagarajah AR (2005) Sustainable wastewater reuse through private sector participation – the Adelaide experience. Retrieved September 2, 2005, from http://www.adb.org/Documents/Events/2005/Sanitation-Wastewater-Management/paper-thiyagarajah.pdf

United Nations Educational, Scientific and Cultural Organisation (2003) Water for people, water for life. The United Nations World Water Development Report-1. Published jointly by UNESCO, France and Berghahn Books, USA

United Nations Educational, Scientific and Cultural Organization (2006) Water – a shared responsibility. The United Nations World Water Development Report-2. Published jointly by UNESCO, France and Berghahn Books, USA

Chapter 2
Urban Wastewater Reuse—A Common Reality

Exponential growth of population, rapid industrialisation and urbanisation, higher cultivation intensities, and poor water management practices over the past century has made freshwater availability a limiting factor in agricultural development (Ray and Gul 1999; Dupont 2003). In addition, the options for increasing supply have become expensive and often environmentally damaging (Frederick 2001). The United Nations World Water Development Report-2 (UNESCO 2006) clearly states that:

> the insufficiency of water is primarily driven by an inefficient supply of services rather than by water shortages. Lack of basic services is often due to mismanagement, corruption, lack of appropriate institutions, bureaucratic inertia and a shortage of new investments in building human capacity, as well as physical infrastructure (p. 45).

The report further states that water crisis rest on how we as individuals, and as part of collective society, govern water resources and their benefits. Therefore, what is needed is managing the available freshwater resources effectively and use them based on fitness-for-purpose criteria. Also, our actions to counter water scarcity challenges should be sustainable, without depleting the natural resources or harming the environment. For these reasons, water managers and policy makers around the world are forced to continually look for alternatives to supplement limited and depleting freshwater resources. In such situations, 'source substitution' appears to be the solution as it allows higher quality water to be reserved for domestic supply and poor quality water may satisfy less critical uses (Hespanhol 1997). Consequently, urban wastewater (treated) is considered as a reliable alternative water source, and wastewater management is assuming prominence in the water management agenda of many countries (Asano 2001; Hespanhol 1997; Cullen 2004).

© The Author(s) 2017
G. Keremane, *Governance of Urban Wastewater Reuse for Agriculture*,
SpringerBriefs in Water Science and Technology,
DOI 10.1007/978-3-319-55056-5_2

2.1 Source Substitution—Response to Freshwater Scarcity Challenge

Wastewater (re)use, particularly for non-potable purposes is an age-old practice (see Table 2.1), and was mainly uncontrolled. Such practices went unreported because the norm then was that wastewater should be treated before use (Ensink et al. 2002). The earliest documented experiment of wastewater use is the large-scale cropland application of municipal wastewater in Western Europe and North America during the early 1900s, when flush toilets and sewer systems were being introduced into these cities (Asano and Levine 1996; Asano 2001; van der Hoek 2004). Since then, there has been an increase in the extent of wastewater usage and applications, and in recent times severe water shortages have pushed the idea of wastewater reclamation and reuse to the forefront of water management discussions.

2.2 Urban Wastewater—Reuse Options and Applications

Urban wastewater reuse may be planned or unplanned; and planned reuse can be direct or indirect. Unplanned reuse is mostly confined to non-potable uses, even though we can find some cases of unplanned potable reuse. Figure 2.1 illustrates the typology of wastewater usage and applications with examples.

As already stated, planned reuse can be direct or indirect. Planned direct reuse can be for potable[1] or non-potable purposes. Planned direct potable reuse is the deliberate use of treated wastewater for some beneficial purpose such as drinking. It is the use of reclaimed water[2] straight from a wastewater treatment plant through a pipe-to-pipe system that connects the reclaimed water line directly to an established potable water supply system without intervening discharge to a natural water body. However, cases of planned direct potable reuse (treated wastewater directly reused for drinking water) are very rare, because of the perception of increased potential risk to public health and because of negative public perception. In general, even though the technology is well proven, direct potable reuse has occurred only when there is no other option, as in the case of Windhoek, Namibia, which is currently the only place where direct potable reuse takes place on a municipal scale.

Planned direct non-potable reuse is the use of treated wastewater where control exists over the conveyance of the wastewater from the point of discharge from a treatment plant to a controlled area where it is used for irrigation. Many countries in

---

[1]An augment of drinking water supplies by highly treated reclaimed water. This includes direct and indirect potable water reuse.

[2]Treated effluent suitable for an intended water reuse application and is synonymous with 'reused water'.

Table 2.1 Examples of water reuse experiments around the world

Year	Location	Purpose/usage
1912–1985	Golden Gate Park, San Francisco, California, USA	Water lawns and supplying ornamental lakes
1926	Grand Canyon National Park, Arizona, USA	Toilet flushing, lawn sprinkling, cooling water, and boiler feed water
1929	City of Pomona, California, USA	Irrigation of lawns and gardens
1942	City of Baltimore, Maryland, USA	Metals cooling and steel processing at the Bethlehem Steel company
1960	City of Colorado Springs, Colorado, USA	Landscape irrigation for golf courses, parks, and freeways
1961	Irvine Ranch Water District, California, USA	Irrigation, industrial and domestic uses, toilet flushing
1962	La Soukra, Tunisia	Irrigation with reclaimed water for citrus plants, reduce saltwater intrusion into groundwater
1968	City of Windhoek, Namibia	Advanced direct wastewater reclamation system to augment potable water supplies
1969	City of Wagga Wagga, Australia	Landscape irrigation of sporting fields, lawns, and cemeteries
1977	Dan region project, Tel-Aviv, Israel	Groundwater recharge and unrestricted crop irrigation
1984	Tokyo Metropolitan Government, Japan	Toilet flushing
1985	City of El Paso, Texas, USA	Groundwater recharge by direct injection into aquifers, and power plant cooling
1987	Monterey regional water pollution control agency, California, USA	Monterey wastewater reclamation study for agriculture—irrigation of food crops
1989	Shoalhaven heads, Australia	Irrigation of gardens and toilet flushing in private residential dwellings
1989	Consorci de la Costa Brava, Girona, Spain	Golf course irrigation
1999	Northern Adelaide Plains, South Australia	Class 'A' water used to irrigate horticulture crops
	Willunga Basin, Adelaide, South Australia	Class 'B' water used to irrigate premium quality grapes
	Mawson Lakes reclaimed water scheme	A dual water reticulation system, recycled wastewater used for toilet flushing, and garden watering

(continued)

Table 2.1 (continued)

Year	Location	Purpose/usage	Year	Location	Purpose/usage
1970	Sappi Pulp and Paper Group, Enstra, South Africa	Industrial use for pulp and paper processes	2000	Rouse Hill recycled water scheme	A dual water supply system, with the recycled wastewater used for toilet flushing, car washing and garden watering
1976	Orange County Water District, California, USA	Groundwater recharge by direct injection into the aquifers	2003	Singapore	The 'NEWater' project provides safe, reliable source of high quality drinking water (sewer water purified to drinking water standards) for Singapore's 4.3 million residents

Source Compiled from Asano (2001), USEPA (2004), Anderson (2003), Salgot and Tapias (2004), van der Hoek (2004) and NEWater (2002)

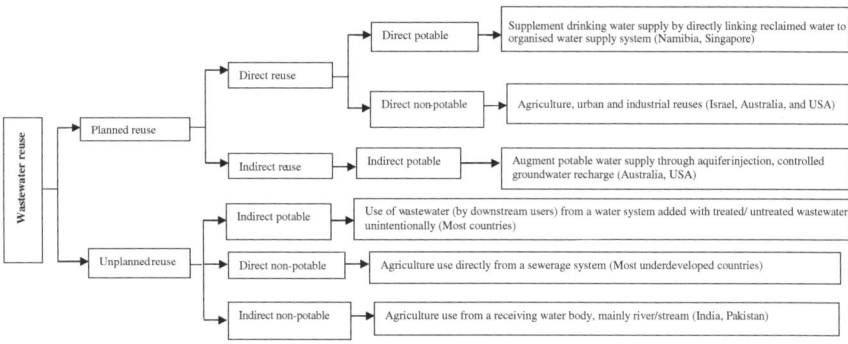

Source: Compiled from Anderson (2003); Salgot and Tapias (2004) and van der Hoek (2004)

Fig. 2.1 Typology of wastewater usage for all purposes

the Middle East and countries like Australia, the United States of America, and Israel have developed large-scale irrigation schemes delivering reclaimed water for agriculture use, using reclaimed water after it has passed through water bodies like storages or wetlands following treatment, or taken from a river, lake, or aquifer that has received sewage or sewage effluent.[3]

Further, planned reuse can also be indirect potable reuse, by way of replenishment of ground water by the controlled addition of reclaimed water to the ground water basin through methods such as aquifer injection. Generally, planned indirect potable reuse is not thought to pose any health risk since it relies on natural treatment in surface water and aquifers, and the reclaimed water is diluted with 'ordinary' river or ground water before extraction, thus ensuring good drinking water quality (WHO 2006). Nevertheless, this is still a new approach and is restricted mostly to the developed countries.

Unplanned reuse, on the other hand is largely for non-potable purposes that can be direct or indirect. Unplanned direct non-potable reuse is the supply of wastewater directly to the land from a sewerage system or other purpose-built wastewater conveyance system. Such situations are found in most of the under developed nations facing water scarcity (Westcot 1997). Unplanned indirect reuse for non-potable purposes is common in developing countries like India and Pakistan, where irrigation water is drawn from rivers or other natural water bodies that receive wastewater flows, treated or not. However, unplanned potable reuse (common worldwide) is also practiced, largely unintentionally, when treated or untreated wastewater is added to a water supply system (reservoirs or rivers or streams) that is subsequently used by downstream communities as a water source for potable use, usually with additional treatment.

---

[3]Water that flows out of treatment plants.

2.3   Urban Wastewater Reuse Experiences

(Re)use of urban wastewater has increased in many places mainly because of increasing demand by the agriculture sector. The best way of using treated wastewater is in agriculture (Pescod 1992), and doing so can definitely relieve a great deal of pressure on fresh water resources. Replacement of freshwater by treated or untreated wastewater is seen as an important conservation strategy contributing to agricultural production. Further, the communities depending on wastewater reuse for their livelihoods, particularly in the developing world, can derive substantial benefits from using nutrient-rich wastewater.

Although wastewater reuse occupies a prominent place in water management policies today, there is no common regulation(s) of wastewater reuse across the world. This is mainly due to different economic and social conditions, and country or state-specific policies towards using wastewater. In developing countries like India, because of increasing water scarcity, lack of money for treatment and a clear willingness by farmers to use untreated wastewater, the practice of using untreated wastewater for irrigation is still being practiced (Ensink et al. 2002). Besides, the technology necessary to produce effluent of a required quality is often unavailable or not maintained, and the regulatory agencies can seldom enforce standards. Nevertheless, some developing countries including India have their own standards adapted from the leading standards set by the FAO or WHO (Achilleos et al. 2005). In developed nations, on the other hand, public health regulations and water pollution control requirements for treatment protect the agricultural workers and the consumers of crops irrigated with treated wastewater. So, wastewater reuse in developing countries is largely unplanned and uncontrolled whereas in developed countries it is controlled and planned (Parkinson and Tayler 2003). In other words, urban wastewater resue in agriculture is 'formal' in developed countries implying that there is some form of fixed irrigation infrastructure, designed and possibly operated by the government or a donor agency, and used by more than one farm household, and in developing countries like India it is 'informal' which menas irrigation is practiced by individuals or groups of farmers, without an irrigation infrastructure planned, constructed or operated by a government or donor agency (Cornish et al. 1999; van der Hoek 2004; IWMI 2007). Hence, use of wastewater for irrigation differs across the world and below are few examples from across the world. These experiences include planned, unplanned, potable, and non-potable applications.

2.3.1   Windhoek, Namibia

The history of wastewater reuse in Namibia dates back to 1968, when the City Council of Windhoek was forced to implement direct reclamation of wastewater for potable use as the city was approaching the limit of its conventional drinking water sources (World Bank 2003). The first water reclamation plant—The Old Goreangab Water Reclamation Plant (OGWRP)—after successful operation for more than 30 years was nearing the end of its viable life in the late 1990s. Therefore, the New Goreangab Water Reclamation Plant (NGWRP) was built in 2002 through a 20-year operation and maintenance (O&M) contract between the City of Windhoek and the Windhoek Goreangab Operating Company Ltd. (WINGOC), which is a consortium of three international water treatment contractors (Lahnsteiner and Lempert 2007). As a result, the city's total water supply is now met by three main sources—(i) surface water, (ii) ground water, and (iii) reclaimed water from both the water reclamation plants.

Initially only 3–8% of reclaimed water was blended with premium water from other sources (bore holes and treated surface water). After several process improvements the portion of reclaimed water was raised gradually until it consti-tuted up to 18% of the total potable water for the city (Lahnsteiner and Lempert 2007). While there is opposition to use recycled water for potable purposes in many parts of the world (Hurlimann and McKay 2006), the people of Windhoek derive pride from the fact that they are the only people in the world with potable reuse. The potable reuse project in Namibia is successful because of the specific attitudes of the users, derived from a growing scarcity of water and a different set of cultural values (McKay 2007a) and a set of water institutional reforms in Namibia, based on proper process design and quality management and on effective public awareness pro-grams (Lahnsteiner and Lempert 2007; Heyns 2005).

2.3.2 Singapore

Singapore like Namibia has achieved a remarkable progress in water resource management as a result of efforts to create a comprehensive management system for the environment, the urban catchment and wastewater. The Four National Taps Strategy has resulted in diversification of Singapore's water sources (World Bank 2006) which includes water from local catchments, imported water (from Malaysia), the NEWater (drinking-quality water produced by treating secondary effluent), and desalinated water. The strategy is a success due to ongoing govern-ment support, institutional integration, integrated land use planning, effective enforcement of legislation, public education, and application of advanced technology.

The water institutions in Singapore provide favourable conditions for Integrated Water Resources Management (IWRM). The administrative barriers facing the process of adopting the IWRM approach that exist in many other countries are largely wiped away in Singapore and it has a comprehensive environmental leg-islation and strict implementation of water resource related regulations.

2.3.3 United States of America

In the United States, urban wastewater management strategies can be categorised as centralised or decentralised. However, from the end of the nineteenth century to the present day, centralised management has remained the preferred urban wastewater management method (Burian et al. 2000). Reclaimed water use in the United States is well established and ranges from pasture irrigation using partially treated reclaimed water to augmenting potable water supplies with highly treated reclaimed water. However, there are no federal regulations governing wastewater reuse, and the regulations and guidelines are developed at the State level (Crook and Surampalli 2005); and therefore, they vary across states. The first regulation of wastewater reuse for irrigation was developed in 1918 (Asano and Levine 1996) and is comprehensive with regard to public health.

2.3.4 Europe

When compared to other regions of the world, Europe has plentiful water resources. However, droughts experienced in the early 90s and in 2003 changed the situation in Europe, resulting in growing water stress, both in terms of quantity and quality (Hochstrat et al. 2005; Bixio et al. 2006). To counter water scarcity challenges the European Union and its member states have enacted the Water Framework Directive (WFD[4]) which highlights an integrated approach to water resources management. Further, the WFD favours municipal wastewater reclamation and reuse to augment water supply and decrease the impact of human activities on the environment (Bixio et al. 2006).

Water reuse is a growing field and many projects have been proceeding throughout Europe in the last fifteen years (Angelakis et al. 2003). Wastewater reuse in Europe is mainly for agriculture, industry, urban, recreational and environmental uses. As compared to the early 1990s, when wastewater reuse in Europe was limited and incidental, at present there are more than 200 fully operational water reuse projects, with many others in an advanced planning phase (Hochstrat et al. 2005). Nevertheless, there are no regulations for wastewater reuse at a European level and the only reference made by the European Union on the matter of wastewater is in the Urban Wastewater Treatment Directive (UWWTD). The UWWTD spells out the implementation of decentralised treatment so as to reduce pollution from households, apply strict sanctions on municipal wastewater treatment plants, and reduce the diffuse pollution from agriculture (Bixio et al. 2006; Achilleos et al. 2005).

---

[4]EU Council Directive for community action in water policy-2000/60/EC of October 23, 2000.

2.3.5   Israel

Due to a combination of severe water shortage, threat of pollution to its water resources and a concentrated urban population with high levels of water consumption and wastewater production, Israel has devoted more effort to wastewater reuse than any other country. Israel's national policy aims to gradually increase the fraction of reclaimed wastewater used instead of fresh water for agricultural use (Brenner et al. 2000). This is reflected by the fact that Israel occupies second place in overall wastewater reuse after California and has the highest percentage of wastewater reused for agricultural irrigation in the world (Achilleos et al. 2005). It is estimated that by the year 2020, 50% of agricultural water consumption will be provided by treated wastewater (Brenner et al. 2000). Understandably, the large-scale wastewater reuse schemes in Israel are mainly for agricultural irrigation. Though modern treatment technology can produce reclaimed water meeting drinking water quality, because of public acceptance considerations the focus in Israel is directed towards maximising saving or replacing freshwater for consumptive uses other than drinking. The Ministry of the Environment determines recommendations for effluent quality standards for various purposes.

The Country specific experiences highlight the potential of this valuable resource and it establishes that different countries have developed different approaches for wastewater reuse to protect public health and the environment. Developed countries have established conventionally low-risk guidelines based on a high technology/high-cost approach, while in developing countries the strategy is to adopt a low technology/low-cost approach based on WHO recommendations (Achilleos et al. 2005). Yet, the objective behind all the guidelines is to achieve better health protection by implementing stringent water quality limits and by defining other appropriate practices, depending on the type of reuse (USEPA 2004).

2.4   Challenges for Wastewater Management

Water resource management in the past was largely shaped by an engineering approach (Pahl-Wostl 2002). However, given the transformation the water sector is undergoing at all levels, water resource management in general is encountering new challenges that call for fresh strategies and institutional arrangements. Likewise, when we think of wastewater management, we find various obstacles. Despite all the potential that (treated) wastewater offers for augmenting freshwater supplies, implementing sustainable wastewater reuse schemes encounters many impediments.

Experiments of wastewater reuse projects around the world suggests that human health, economic prosperity, property rights, and a general responsibility to the natural environment are the important components of accomplishing effective wastewater solutions (Jones 2005). Furthermore, the multidimensional character of this resource—time, space, multidiscipline, and stakeholders-make it important to consider a large number of parameters in the decision making processes. These include: 'sustainability issues, legislation and health issues, techniques and technology, political and institutional issues, socio-economic impacts, and historical and cultural issues' (Thomas and Durham 2003, p. 24). According to Livingston et al. (2004, p. 581), 'successful implementation of new approaches to wastewater management is a multi-faceted challenge requiring input beyond mere technical'. Societal and institutional adaptation is therefore critical to ensuring long-term sustainability of reuse schemes (Asano 2001; Po et al. 2004, 2005; Mills 2000, Kasower 1998, Ritchie et al. 1998, all cited in Haddad 2002; Livingston et al. 2004).

2.4.1 Institutional Challenges

Wastewater collection, treatment and effluent use normally encompass a wide range of interests at different levels of administration. So the scope and success of any reuse scheme will depend to a large extent on the institutional organization (Pescod 1992). In any natural resource management regime, coordination complexity results in problems, due to varying roles and responsibilities and overlapping concerns among the public agencies managing the resources (MacDonald and Dyack 2004; McKay 2007b). Previous studies related to wastewater use (Asano 2001; Po et al. 2004, 2005) have identified similar conflicting agendas among water agencies: addressing water rights issues; dealing with opponents to recycling or reuse; modifying existing regulations; and acquiring funding, are the institutional challenges facing successful development of this dependable resource. Therefore, appropriate institutions with adequate resources are required for development of sustainable wastewater reuse schemes. More about institutions is presented in Chapter Four.

2.4.2 Public Perceptions and Acceptance

For successful implementation of reuse schemes, public acceptance is a very important (Asano 2001; Po et al. 2004; Marks 2004; Marks et al. 2006; McKay and Hurlimann 2003). Generally, the tendency of people to be motivated by a set of long-term goals, but to act in the short term towards those things that they control, is what affects wastewater reuse projects (Jones 2005). Therefore, understanding public perceptions and community acceptance of water reuse is very important. Failure to gain public acceptance has led to vocal opposition and, at times, has resulted in schemes being stalled. According to Robinson et al. (2005), public concerns about real or perceived risks are weighted against the use of reclaimed water.

There are very few studies that have tried to investigate the factors influencing public perceptions of water reuse and their influence on individuals' decision-making processes. It is only in the recent past that public perceptions and acceptance of water reuse have been considered important for successful implementation of reuse schemes. While reviewing the existing international and Australian literature on water reuse, Po et al. (2004) have identified the following factors to influence community's acceptance of a reuse scheme: disgust or 'yuck' factor, the perception of risks associated with using recycled water, the specific uses and cost of recycled water, the sources of water to be recycled, issues of choice, trust and knowledge, attitudes toward the environment, and socio-demographic factors. If wastewater resources are to become an integral component of water and waste management policies, the acceptance of reclaimed water must be comprehensively tackled; this is more critical if the application is for potable uses. However, this challenge can be systematically addressed through effective educational, policy, and management strategies, as in case of the Windhoek Water Reclamation Project, Namibia or the NEWater in Singapore.

Wastewater reuse history is marked with failure of reuse schemes mainly because of lack of community involvement (Po et al. 2004, 2005; Hurlimann and McKay 2006). According to Jones (2005), 'working with a community that does not have wastewater as a highest priority requires building participation through a combination of discussions about community outcomes, and more detailed action steps of technology identification, design work, and management'. Since it is the public who will be served by and pay for them, the policies on wastewater use and management must include the human dimension (Robinson et al. 2005). For a reuse scheme to be sustainable, community involvement and/or participation are very important. Asano (2001) suggests that water reuse project(s) should be built upon three principles:

- providing reliable treatment of wastewater to meet strict water quality requirements,
- protecting public health, and
- gaining public acceptance.

2.4.3   Market Imbalance

The best application for the use of wastewater after treatment is in agriculture (Pescod 1992) and use of this water for agriculture purposes can relieve a great deal of pressure on fresh water resources. This implies that the largest market for reclaimed water is in the agriculture sector. In addition, use of wastewater, mainly for non-potable purposes is also increasing. Although there is market for this valuable resource, it is imbalanced, as is explained by Abu Madi et al. (2003, p. 115):

the market for reclaimed water is unbalanced and it is due to a growth on the supply side of the market, revealed by increasing number of wastewater treatment plants and stagnancy on the demand side revealed by the substantial proportions of the resource being discharged without proper utilization.

The reason for this imbalance is once again the institutional challenges facing implementation of reuse schemes and lack of community involvement in the implementation of those schemes. Success or failure of reclaimed water schemes largely depend on institutional factors, such as federal and/or state financial support, devolution of sufficient authority to local authorities, and development of innovative resource management institutions (Mills 2000; Kasower 1998; Ritchie et al. 1998, all cited in Haddad 2002).

2.4.4 Financial Feasibility and Technicality

In addition to the above-mentioned impediments, financial feasibility is also important while implementing water reuse projects. Financing a reuse scheme is a challenge because acquiring funds to develop a water reuse scheme is an onerous task. Users' willingness to pay for the resource in question (wastewater in this case) to a large extent also influences the implementation of reuse schemes. According to Tsagarakis and Georgantzís (2003, p. 112), 'more often than is usually believed, individually rational behaviour is compatible with socially desirable outcomes'. Therefore, public perceptions and acceptance of wastewater, community participation and willingness to pay are all interlinked.

Willingness to pay for reclaimed water is also influenced by the tariff structure adopted in a particular scheme. The general tendency observed in case of water reuse schemes is that users might not be willing to pay more for this resource because it is considered as waste, so why pay for it? Therefore, the tariff structure should be such that the community being served should perceive it to be appropriate, as well as taking into account the long term viability of the service provider.

Sound technicality is another factor to be considered while implementing reuse projects. This is important because the effluent should be treated to a quality acceptable to the end user and matched to particular application. In the present context reclaimed water used for agricultural irrigation must be of very high quality to meet the process needs of the agriculture industry and to minimize the potential impacts on human health by inadvertent exposures. Therefore, the acceptability of reclaimed water for different uses is dependent on the specific application and is highly variable. In developing countries like India, treatment facilities are either not available or not implemented on the grounds of cost. In such situations, farmers using wastewater should be encouraged to adopt safer approaches. This can be achieved through participatory approaches such as farmer's field schools and public health education.

References

Abu Madi M, Braadbaart O, Al-Sa'ed R, Alaerts G (2003) Willingness of farmers to pay for reclaimed wastewater in Jordan and Tunisia. Water Sci Technol: Water Supply 3(4):115–122

Achilleos A, Kythreotou N, Fatta D (2005) Development of tools and guidelines for the promotion of the sustainable urban wastewater treatment and reuse in the agricultural production in the Mediterranean Countries. Task 5: Technical guidelines on wastewater utilisation, European Commission, June 2005

Anderson J (2003) The environmental benefits of water recycling and reuse. Water science and technology: water supply, 3(4):1–10

Angelakis AN, Bontoux N, Lazarova V (2003) Challenges and prospective of water recycling and reuse in European countries. Water Sci Technol: Water Supply 3(4):59–68

Asano T (2001) Water from (waste) water—the dependable water resource. Paper presented at the 11th Stockholm Water Symposium, Stockholm, Sweden, 12–18 Aug 2001

Asano T, Levine AD (1996) Wastewater reclamation and reuse: past, present, and future. Water Sci Technol 33(10–11):1–14

Bixio D, Thoeye C, De Koning J, Joksimovic D, Savic D, Wintgens T, Melin T (2006) Wastewater reuse in Europe. Desalination 187:89–101

Brenner A, Shandalov S, Messalem R, Yakirevich A, Oron G, Rebhun M (2000) Wastewater reclamation for agricultural reuse in Israel: trends and experimental results. Water Air Soil Pollut 123:167–182

Burian SJ, Nix SJ, Pitt RE, Rocky Durrans S (2000) Urban wastewater management in the United States: past, present, and future. J Urban Technol 7(3):33–62

Cornish GA, Mensah E, Ghesquière P (1999) Water quality and peri-urban irrigation. An assessment of surface water quality for irrigation and its implications for human health in the Peri-urban zone of Kumasi, Ghana. (Report OD/TN 95). HR Wallingford Ltd., Wallingford, UK

Cullen P (2004) Water challenges for south Australia in the 21st century. Adelaide thinker in residence, Department of the premier and cabinet, p 11

Crook J, Surampalli RY (2005) Water reuse in the United States. Water Sci Technol Water Supply 5(3–4):1–7

Dupont A (2003) Will there be water wars? Development Bulletin, 63 Nov 2003:16–20

Ensink JHJ, van der Hoek W, Matsuno Y, Munir S, Aslam RM (2002) Use of untreated wastewater in peri-urban agriculture in Pakistan: risks and opportunities, (Research Report-67). International Water Management Institute, Colombo, Sri Lanka

Frederick KD (2001) Water marketing: obstacles and opportunities. Forum for Appl Res Public Pol 16(1):54–62

Haddad BM (2002) Monterey county water recycling project: institutional study. J Water Res Planning and Manage 128(4):280–287

Hespanhol I (1997) Wastewater as a resource. In: Helmer R and Hespanhol I (Eds) Water pollution control: a guide to use of water quality management principles, published by E and F. Spon on behalf of WHO/UNEP

Heyns P (2005) Water institutional reforms in Namibia. Water Policy 7:89–106

Hochstrat R, Wintgens T, Mellin T, Jeffrey P (2005) Wastewater reclamation and reuse in Europe: a model-based potential estimation. Water Sci Technol: Water Supply 5(1):67–75

Hurlimann A, McKay JM (2006) What attributes of recycled water make it fit for residential purposes? The Mawson Lakes experience. Desalination 187:167–177

International Water Management Institute (2007) IWRM challenges in developing countries: lessons from India and elsewhere. Water Policy Briefing (24), Jan 2007

Jones K (2005) Engaging community members in wastewater discussions. EcoEng Newsletter (11), Oct 2005

Lahnsteiner J, Lempert G (2007) Water management in Windhoek, Namibia. Water Sci Technol 55(1–2):441–448

Livingston D, Stenekes N, Colebatch HK, Ashbolt NJ, Waite TD (2004) Water recycling and decentralised management: the policy and organisational challenges for innovative approaches. In: Daniell T (ed) Proceedings of the International Conference on Water Sensitive Urban Design: Cities as Catchments, Adelaide, 21–25 Nov 2004:581–592

MacDonald DH, Dyack B (2004) Exploring the institutional impediments to conservation and water reuse–National issues. CSIRO Land and Water Client Report

Marks JS (2004) Advancing community acceptance of reclaimed water. Water 31(5):46–51

Marks JS, Martin B, Zadoroznyj M (2006) Acceptance of water recycling in Australia: national baseline data. Water 33(2):151–157

McKay JM (2007a) Policy changes in South Australia: elements of the social contract resulting in high urban and agricultural use of recycled water. South Australia Policy Online. Retrieved June 13 2007 from http://www.sapo.org.au/opin/

McKay JM (2007b) Water governance regimes in Australia: implementing the National water initiative. Water 34(1):150–156

McKay JM, Hurlimann A (2003) Attitudes to reclaimed water for domestic use: part 1. Water 30 (5):45–49

NEWater (2002) NEWater sustainable water supply. Retrieved May 10 2007, from http://www.pub.gov.sg/NEWater

Pahl-Wostl C (2002) Participative and stakeholder-based policy design, evaluation and modelling processes. Integr Assess 3(1):3–14

Parkinson J, Tayler K (2003) Decentralised wastewater management in peri-urban areas in low-income countries. Environ Urbanisation 15(1):75–89

Pescod MB (1992) Wastewater treatment and use in agriculture. Irrigation and Drainage (Paper No. 47). FAO, p 118

Po M, Juliane K, Nancarrow BE (2004) Literature review of factors influencing public perceptions of water reuse. Australian Water Conservation and Reuse Research Program, CSIRO

Po M, Nancarrow BE, Leviston Z, Porter NB, Syme GJ, Karecher JD (2005) Predicting community behaviour in relation to wastewater reuse: what drives decisions to accept or reject? Water for healthy country national research flagship. CSIRO Land and Water, Perth

Ray I, Gul S (1999) More from less policy options and farmer choice under water scarcity. Irrigation and Drainage System, 13:361–383

Robinson KG, Robinson CH, Hawkins SA (2005) Assessment of public perception regarding wastewater reuse. Water Sci Technol: Water Supply 5(1):59–65

Salgot M, Tapias JC (2004) Non-conventional water resources in coastal areas: a review on the use of reclaimed water. Geologica Acta, 2(2):121–133

Thomas J, Durham B (2003) Integrated water resource management: looking at the whole picture. Desalination 156:21–28

Tsagarakis KP, Georgantzís N (2003) The role of information on farmers' willingness to use recycled water for irrigation. Water Sci Technol: Water Supply 3(4):105–113

United States Environmental Protection Agency (2004) Guidelines for water reuse. United States E.P.A. for International Development, Washington, DC

van der Hoek W (2004) A framework for a Global assessment of the extent of wastewater irrigation: the need for a commn wastewater typology. In: Scott C, Faruqui NI, Raschid L (eds) Wastewater use in irrigated agriculture–confronting the livelihoods and environmental realities. CABI/IWMI/IDRC

World Bank (2003) Wastewater reuse. Water Resources and Environment Technical Note F.3. The World Bank, Washington, DC

World Bank (2006) Dealing with water scarcity in Singapore: institutions, strategies, and enforcement. China, Addressing Water Scarcity. (Background paper No. 4). World Bank Analytical and Advisory Assistance Program. Washington, DC

Chapter 3
Water Governance and Wastewater Reuse in Australia and India

Organising the water sector is largely influenced by a country's overall standard of governance, its customs, politics and conditions (Rogers and Hall 2003; UNESCO 2006) resulting in variations in the ways the water sector is organised around the world. For many years, 'governance' was discussed and debated extensively in the context of society and development as a whole. But, in recent times, because water crisis is observed as 'a crisis of governance' the notion of good governance has attracted lot of attention from water managers, planners, and policy makers (Rogers and Hall 2003).

3.1 Water Governance—Concept and Definitions

Governance and management are interdependent in the sense that effective governance systems should enable practical management tools to be applied correctly (UNESCO 2003). Generally, the terms governance and government are used as synonyms, but in reality they differ.

The phrase 'governance' can be used in several contexts; characterised in a number of ways; and has a range of definitions. Accordingly, it is presented in many forms in the development literature: 'global governance' (Keohane 2003), 'self-governance' (Ostrom 1990; Tang 1992), 'modern governance' (Gaudin 1998), 'water governance' (Rogers and Hall 2003), 'distributed governance' (Townsend and Pooley 1995). Below are some definitions of governance.

The UNDP (2006) defines governance as:

an exercise of economic, political and administrative authority to manage a country's affairs at all levels. It comprises the mechanisms, processes and institutions through which citizens and groups articulate their interests, exercise their legal rights, meet their obligations and mediate their differences (pp. 35–36).

© The Author(s) 2017
G. Keremane, *Governance of Urban Wastewater Reuse for Agriculture*,
SpringerBriefs in Water Science and Technology,
DOI 10.1007/978-3-319-55056-5_3

Rogers and Hall (2003), p. 4 distinguish the terms government and governance and define governance as:

a more inclusive concept than government per se; it embraces the relationship between a society and its government. Governance generally involves mediating behaviour via values, norms, and, where possible, through laws......it also relates to government policies and actions.

According to Stoker (1998), governance is ultimately concerned with creating the conditions for ordered rule and collective action while Cleaver and Franks (2005) argue that governance is usually seen to entail 'doing things right' which is not true within the water resources sector. As a result, 'good governance'/'effective governance' is the new mantra within the water sector.

3.1.1 Attributes of Good Governance

There is no single definition for good or effective governance but a review of the development literature helps us to identify certain characteristics of good governance (see Table 3.1).

Table 3.1 Attributes representing the features of good governance

Attributes	Features
Participation	All citizens, both men and women, should have a voice—directly or through intermediate organizations representing their interests— throughout processes of policy and decision-making. Broad participation hinges upon national and local governments following an inclusive approach
Transparency	Information should flow freely within a society. The various processes and decisions should be transparent and open for scrutiny by the public
Equity	All groups in society, both men and women, should have opportunities to improve their well-being
Accountability	Governments, the private sector and civil society organizations should be accountable to the public or the interests they are representing
Coherency	The increasing complexity of water resource issues, appropriate policies and actions must be taken into account so that they become coherent, consistent and easily understood
Responsiveness	Institutions and processes should serve all stakeholders and respond properly to changes in demand and preferences, or other new circumstances
Integrative	Water governance should enhance and promote integrated and holistic approaches
Ethical considerations	Water governance has to be based on the ethical principles of the societies, in which it functions, for example by respecting traditional water rights

Source UNESCO (2003, p. 373)

Grindle (2002) structures good governance around six principles: participation, fairness, decency, accountability, transparency, and efficiency. Similarly, the World Bank, Asian Development Bank (ADB), Global Water Partnership (GWP), United Nations Economic and Social Commission for Asia and the Pacific (UNESCAP) have all identified similar characteristics of good governance. Some of these characteristics, such as open and transparent, inclusive and communicative, equitable and ethical, are related to the approaches used by a governance system. The others, like accountability, efficient, responsive and sustainable are related to its performance and operation (Rogers 2002; Rogers and Hall 2003).

While these attributes are of an ideal model, in real world it is difficult to achieve good governance in its totality. Yet, to ensure sustainable human development, actions must be taken to work towards the ideal model with the aim of making it a reality. In relation to water governance, this can be achieved only when the institutions produce results that meet the needs of society by making the best use of resources at their disposal, thus leading to the sustainable use of natural resources and the protection of the environment.

3.1.2 Water Governance

The UN World Water Development Report-2 (2006), while reporting on the state of global water governance observed that:

in many countries water governance is in a state of confusion: in some countries there is a total lack of water institutions, and others display fragmented institutional structures or conflicting decision-making structures (UNESCO 2006, p. 44).

Similarly, the World Panel on financing global water infrastructure, in its report Financing Water for All (Winpenny 2003), reported that serious defects in the governance of the global water sector are at the root of all the problems. For these reasons, water planners and policy makers agree that governance is one of the biggest challenges within the water sector.

Within the water sector, the concept of governance is commonly used as a synonym for management, defined as the collective allocation of resources to achieve specific objectives (Cleaver and Franks 2005). The authors further suggest that management implies managers interacting with stakeholders in the process of achieving outcomes, while governance describes the interactions between stakeholders to achieve them. Thus, water governance is concerned with the ongoing processes of extracting, distributing and using water within the present institutions. According to Rogers and Hall (2003),

Water governance is a range of political, social, economic and administrative systems that are in place to develop and manage water resources, and the delivery of water services, at different levels of society (p. 12).

Water governance therefore is a framework of political, social, economic, and legal structures within which societies choose and accept to manage their water related affairs. It includes governments, the market forces that help to allocate resources, and any other mechanisms that regulate human interactions. It can be looked upon as processes of decision-making, involving both formal and informal actors in society at all levels—government is just one of these actors—and based on the outcomes of these processes, governance can be 'good' or 'bad'. For a governance system to be good or effective it should exhibit certain characteristics (discussed earlier) that are often difficult to achieve completely.

3.2 Water Transition in Australia and India

With the global population growing rapidly, rapid industrialisation and urbanisation, the traditional water systems around the world are coming under pressure (Dirksen 2002). In developing countries such as India, the dominance of agricultural water use is the driving force for policy reforms, while in the industrialised countries like Australia, it is urban, industrial and environmental water demands that have spurred them (Turrall 1998). As a result of reforms there has been a paradigm shift in water resources management policy around the world (Gleick 2000). Also, the ways of organising the water sector vary across countries, since they reflect local political, cultural, and administrative traditions (Rogers and Hall 2003). Given the variations in water resources management, there is consensus among most researchers and policy makers in the water sector about making a transition from a water resource development mode to a water resource management mode, by embracing Integrated Water Resources management (IWRM) (Shah and van Koppen 2006). Consequently, the way in which water is managed has changed considerably over the years due to continuously evolving technologies, altered understandings and perceptions of water, new lifestyles, and economic development (Huitema and Meijerink 2007). Water governance today involves a range of stakeholders-the government, civil societies, and the private sector, each with their own responsibilities. It is concerned with how institutions rule and how regulations affect political action and the prospect of solving societal problems, such as efficient and equitable allocation of water resources (UNESCO 2003, p. 372). Therefore, understanding the dynamics surrounding the development, introduction, and implementation of such institutional change, and how it occurs is imperative.

3.2.1 Institutional Change and Water Transition

Institutional change focuses more on the rules and processes that govern relationships between organizations and the public, and different organizations (North

1990; Ostrom 1990). Hargrave and van de Ven (2006), p. 866 define institutional change as 'a difference in form, quality, or state over time in an institution'. According to Hobley and Shields (2000), p. 15, 'institutional change refers to change in the architecture and relationships between agencies and organizations. It addresses issues in the wider environment such as policy, laws, governance structures as well as issues of co-ordination between agencies (for example, contracting-out services)'. But what brings about this (institutional) change?

Baez and Abolafia (2002), based on their data and readings of institutional change literature suggest that:

> Institutional change in organizations rests on three assumptions. First, organizational actors make sense of, or interpret their organizations and environmental contexts in order to simplify the world they live in.... Second, environmental pressures change in sometimes unpredictable and unexpected ways, and actors are affected by these shifts.... Third, the degree to which actors take their context for granted varies with environmental pressure... (p. 527).

Lin (1989) argues that institutional change can be induced or imposed and explains these two different types of institutional change as follows:

> An induced institutional change refers to a modification or replacement of an existing institutional arrangement or the emergence of a new institutional arrangement that is voluntarily initiated, organised, and executed by an individual or a group of individuals in response to profitable opportunities. An imposed change, in contrast, is introduced and executed by governmental orders or laws (Lin 1989, p. 13).

Furthermore, in the New Institutional Economics (NIE) literature we come across two different approaches—demand and supply induced—to explain institutional change (see Wegerich 2001). Like many water scarce regions around the world Australia and India too have had a paradigm shift or transition in the way they manage/govern their water resources.

3.2.1.1 Australia

Australian water sector reforms were the result of a broader reform agenda initiated during the 1980s and 1990s (Srivastava 2004). Between 1960 and 1992, Australia slipped from being the third richest developed nation in the world to the fifteenth position. This drove successive governments to initiate a package of reforms, particularly infrastructure reforms, including the water infrastructure. Since 1992, the Australian Government has embarked on two phases of ambitious reform of state laws and policies for water management. The first, in 1994, is known as the Council of Australian Government (CoAG) reforms, and the second, in 2004, is known as the National Water Initiative reforms (McKay 2006, p. 115).

These water sector reforms were prompted by a number of domestic environmental and social issues and international processes, and were targeted at reducing government activity in water management. However, initiation of the reforms in the water, gas, electricity, and transport industries that were adopted in the form of the National Competition Policy (NCP) in 1995 is seen as the driver of change in the Australian water industry. The water sector reforms in Australia can be explained as follows:

> The NCP and the CoAG Water Reform Agenda are the two principal pillars of government policy stimulating reform in the water industry at the national level. The National Water Quality Management Strategy (NWQMS), which provides guidelines to regulate issues related to public health and the environment, and the National Environment Protection Council (NEPC) are the two other elements of the reform framework (Srivastava 2004, p. 3).

Nevertheless, the purpose of the reforms was to achieve efficient and customer-oriented service by restructuring the public water utilities (McKay and Halanaik 2003). McKay explains these reforms or shifts as 'four paradigms of formal water resources laws and policies since 1788' (McKay 2006, p. 115). As a result of these reforms, every State in Australia has introduced its reforms in a different way, and consequently the water services industry in Australia provides examples of a variety of models for water service provision, a variety of regulatory regimes, and some examples of private sector participation.

Institutional arrangements and regulatory regimes

Across Australia, South Australia (SA), Western Australia (WA) and the Northern Territory (NT), and the Australian Capital Territory (ACT) each have a single state-owned utility with the primary responsibility for water supply and sewerage services. The local governments or local boards are vested primarily with the responsibility for water and sewerage services in New South Wales (NSW), Queensland (QLD), and Tasmania (TAS). The state of Victoria (excluding Melbourne) offers the only example of regional utility model in which more than one utility exists and each of them services multiple local-government agencies. However, this is a very recent evolution.

With respect to ownership and operations, State or local governments own all the water utilities in Australia. With the exception of some irrigation schemes, there has been little privatization in the water sector. However, there has been restructuring and institutional role separation within the public sector departments. The public sector departments have been transformed to corporations, subject to the same laws that govern the private sector, and with clear commercial objectives (Srivastava 2004). Further, a number of water utilities have contracted out their design, construction, and various operational roles to the private sector through service or management contracts. This process is usually termed corporatization wherein government owns the assets but contracts out the management (McKay and Halanaik 2003). This has been achieved through various models available for private sector participation, which will be discussed in Chapter Four. Likewise, Australia has a variety of regulatory regimes: health regulation, environmental regulation and economic regulation.

An economic regulator has the responsibility both for prices and for customer service standards. The emerging trends and practices in Australia with respect to economic regulation show a clear shift towards independent regulation, and most of the States and territory jurisdictions favour a multi-sector approach. For health regulation, in almost all the states the health department controls compliance with national water and sewerage quality standards. Environment regulation comes under an Environment Protection Authority/Agency (EPA) in all states, except in Western Australia and the Northern Territory, where it is the responsibility of a department. Proper pricing of rural and urban water is one of the key issues for reform in the Australian water industry; as a part of the COAG reforms, the 'pay for use' principle was adopted, which provides for water services to earn fair rate of return, ensuring that their business is financially viable and sustainable. All states have adopted a two-part tariff for water provision, consisting of a fixed access fee and a charge for usage. Sewerage charges are generally fixed.

3.2.1.2 Water Transitions in India

As with water transitions all over the world, India too has experienced a water transition; the current water management regime advocates community participation. There has been a major shift in water management, where the Government of India (GoI), through its Sector Reform Programme (SRP), aims to create a sense of ownership and control by local communities of assets created through partial contributions (Joshi 2004). This is true for other programmes for managing the commons, such as participatory watershed development, participatory irrigation management (PIM), and joint forest management (JFM). Although these programmes have been initiated in response to the incapacity of public sector to effectively operate and maintain the resource systems (Nicol 2000, cited in Joshi 2004) and the failure of supply-driven approaches to deliver these services to the rural poor, the major thrust for institutionalising and implementing these programmes has been from the World Bank, through its various projects (Joshi 2004). Although, under the Indian Constitution, provision of water is the responsibility of the State governments, the Union Ministry of Water Resources, at the central level, is responsible for development, conservation and management of water as a national resource. It also oversees the regulation and development of inter-state rivers through various Central organizations. Urban water supply and sewage disposal is the responsibility of the Ministry of Urban Development, while rural water supply is handled by the Department of Drinking Water, under the Ministry of Rural Development. Hydroelectric power is the responsibility of the Ministry of Power, while pollution and environment control comes under the Ministry of Environment and Forests.

Institutional Arrangements

As mentioned earlier, since water is a state matter, the State governments have primary responsibility for its use and control. At the State level, major and medium-sized irrigation projects are handled by irrigation departments, while minor irrigation is looked after partly by water resources departments, minor irrigation corporations, Zilla Panchayats, and other departments such as agriculture. Urban water supply is the responsibility of Public Health Departments and rural water supply is taken care of by Panchayats. Hydropower is the responsibility of the State Electricity Boards.

Adopting Rotmans et al. (2001) idea of transition, in India, water sector has yet to reach the acceleration phase. It can be said that, with the passing of National Water Policy 2001, India has entered the take-off phase, but more needs to be done to move towards the stabilisation phase.

Focusing on the discussions about water governance and water transitions in Australia and India, it is clear that the water governance regime and institutional environment are different in both countries. The difference can be attributed to varying social, economic, and political settings. Moreover, both countries have different water economies in the sense that water economy in Australia is 'formal' while in India the water economy is 'informal'.

The distinction between a formal and informal water economy is based upon the stage of 'formalisation' of the water economy in a particular country which means 'the proportion of the economy that comes under the extent of direct regulatory influence' (IWMI 2007, p. 2). According to (Fiege 1990, cited in Shah and van Koppen 2006), an informal economy is that part of the economy that remains outside formal mechanisms of governance-law, policy and administration. Shah and van Koppen (2006) categorise the water economies around the world into four stages: (1) Completely Informal; (2) Largely Informal; (3) Formalizing; and (4) Highly Formal. The authors focus more on the dominant mode of water service provision and related institutional arrangements. In developing countries like India water users depend largely on self-supply, informal exchanges and local community institutions while in developed countries like Australia most users are served by public or private service providers (IWMI 2007; Shah and van Koppen 2006). As explained earlier, with respect to waste water reuse, Australia has formal arrangements to govern wastewater reuse in agriculture and other purposes and in India it is uncontrolled, and unregulated.

3.3 Wastewater Reuse in Australia and India

The search for a reliable alternative source of water has triggered the development of water reclamation and reuse projects around the world. In developed countries like Australia wastewater use is more controlled and planned whereas in the developing world it is uncontrolled and still done in the de facto illegal manner.

3.3.1   Australia

The scope for Australia to recycle water was first identified during 1977–78 in a report commissioned for the Victorian Government on the potential for water recycling (GHD 1978). However, this failed to attract the attention of the policy makers and hence had little impact until the 1980s, when issues of environmental health, sustainability, water availability and water quality for consumptive uses emerged as significant political issues (Taylor and Dalton 2003). Furthermore, Australia is currently experiencing the highest ever amount of pressure on its water resources. Additionally, it has been stated that "substitution of water used in agriculture and urban irrigation with reclaimed water will free up water and help make appropriate allocations to the environment, thus ensuring good environmental condition for stressed water supplies" (Hamilton et al. 2005, p. 185). This means that reclaimed water can definitely become a major resource for the agriculture sector, because the agriculture industry was the largest consumer of water, consuming 11,814GL of water in 2013–14 (ABS 2016).

Water recycling was given impetus, starting in the early 1990s when the States established Environment Protection Authorities (EPA) which imposed compositional standards on the discharge of treated effluents from sewage treatment plants (STPs) to the oceans. Water recycling was brought within the National Water Reform Framework in 2003. This framework is an intergovernmental agreement aimed to encourage water conservation in cities through better use of storm water and recycled water (Hurlimann and McKay 2006). The subsequent signing of the Intergovernmental Agreement on the National Water Initiative (NWI), and the creation of the Australian Government Water Fund, laid the foundation for encouraging innovation and the use of recycled water in Australia's cities and towns. Thus, a series of events in the late 90s provided powerful incentives for cities and town to consider including water recycling in their water development plans and ultimately converged to accelerate implementation of water recycling (Dillon 2000; Dimitriadis 2005). As a result, the interest increased in recycling for productive purposes on land as an alternative to installing expensive biological nutrient removal plants. The droughts of 2001–3 reinforced the need for more effective water management, with recycled wastewater, urban storm water and rainwater being seen as resources rather than problems.

Most of the wastewater reuse schemes in Australia are irrigation schemes and examples of formal wastewater reuse schemes. The schemes are formal in the sense that the arrangements between different parties/stakeholders (governments, local councils, and private companies) involved in implementing these schemes are formal and clear. Box 3.1 is an example of a formal wastewater irrigation scheme operating in Australia.

Box 3.1 Brighton Irrigation Scheme, Hobart

Brighton is a dry area near Hobart, with significant broad acre farming. In an attempt to drought proof the area, Brighton Council is now recycling all its wastewater (800–900 ML per year), which is reclaimed from residential properties connected to sewerage in the areas Old Beach, Gagebrook, Bridgewater, and the Brighton township itself.

The Brighton Lagoon recycling system was established in 1996 and since then treated effluent has been used for irrigation of poppies, cereals and pasture on the neighbouring farm. In 1997, the Council formed a joint venture with the local pulp mill at Boyer to establish an irrigated pine plantation of 17 hectares. The success of the Brighton Lagoon recycling system enabled Council and local farmers to secure funding to establish infrastructure required to also recycle treated effluent from the Green Point Waste Water Treatment Plant. Participating farmers agreed to install and pay for suitable storage and irrigation infrastructure. The Brighton Council, with the help of a NHT Coasts and Clean Seas program grant of $788,000, paid for the recycled water distribution network. Farmers were irrigating with recycling water twelve months after project funding had been announced, which led to an initial reduction of the demand for potable water by 20%. Apart from the recycled water being an important and affordable source of water to local farmers, the scheme also led to a significantly reduced discharge of nutrients (nitrogen and phosphorus) into the river Derwent, thereby, helping farmers save on fertiliser.

Sources Radcliffe (2004); naiadTM

In addition to wastewater irrigation schemes, Australia has successful examples of dual distribution systems and planned potable reuse practices, usually referred to as Aquifer Storage and Recovery (ASR). A dual distribution system is a situation where there are two water supply lines: one for potable water and another for reclaimed water. Aquifer Storage and Recovery (ASR) is a method of enhancing water recharge to underground aquifers by gravity feeding or pumping excess water into the aquifers for later use in times of peak demand, using excess surface water, including urban storm water runoff, and treated wastewater (Dillon et al. 1999; Martin and Dillon 2002).

3.3.2   India

In India, there is a long history of wastewater use (untreated or partially treated). For ages, the marginalised communities in India have relied on the indirect use of wastewater to grow vegetables, fruits, cereals, flowers and fodder (van der Hoek et al. 2002). In recent years, as a result of rapid population growth, massive industrialization, and the growing number of cities that dispose of large amounts of sewage into bodies of water, the indirect use of wastewater has increased even further. Most wastewater irrigation, in the peri-urban and rural areas of India, occurs along the rivers that flow through such rapidly growing cities. According to UNDP's World Water Development Report (2003), 70% of industrial wastes in developing countries are dumped into waters without treatment, polluting the usable water supply.

Unlike in the developed world, where wastewater irrigation is controlled and carefully planned, in many parts of the developing world wastewater use is indirect and unregulated, which means that the wastewater is disposed of in rivers from where the contaminated river water is then used for irrigation (van der Hoek et al. 2002). The common practice observed is that, untreated urban wastewater is used downstream for uncontrolled, unrestricted irrigation. The water from the rivers that receive wastewater flows is diverted via anicuts (weirs) to canals and often to tanks, and then channelled to the fields for irrigation (Buechler and Devi 2003). Accordingly, most wastewater irrigation in India occurs along rivers, and if it was not for these continuous wastewater flows, many of the rivers of the Indian peninsula would have run dry throughout the year. In some other cases many people irrigate their crops by extracting the wastewater from the nallas (open drains) or the underground sewer pipes (Juwarkar et al. 1988). Box 3.2 presents a case of unregulated/informal wastewater reuse. These practices are more common in the semi-arid regions where the monsoon rains are erratic and unreliable, and hence wastewater is a valuable resource for farmers.

Box 3.2 Unregulated Irrigation with Wastewater in Hubli-Dharwad, Karnataka, India

Within the twin city of Hubli-Dharwad 60 million litres of wastewater is generated every day which flows untreated into the natural watercourses. In the semi-arid climate where the monsoon rains are erratic and unreliable, wastewater is a valuable resource for farmers. Many extract it from the nallas (open drains) and underground sewer pipes to irrigate their crops, and this is considerably cheaper than constructing a borehole and hence the practice is more accessible and attractive to small farmers. Wastewater also provides an irrigation source during the dry season, enabling farmers to sell their produce for five times the monsoon prices, while its high nutrient load reduces the need for costly fertilizer inputs.

While this practice alleviates poverty for many farmers, it simultaneously places them, the consumers, and environment at risk. Untreated wastewater is a major source of pathogens, water-borne parasites and also contains potentially injurious bio-medical waste (including disposable needles and syringes), creating serious health concerns. Continuous irrigation with wastewater also leads to environmental problems such as salinisation, phytotoxicity (plant poisoning) and soil structure de-teriotation, which in India is commonly referred to as 'sewage sickness'. Therefore, such wastewater irrigation practices reveal a range of associated problems that threaten to outweigh the benefits.

Source Bredford et al. (2003)

Over the past two decades wastewater use in agriculture has increased significantly. And with the growing population and increased industrial use of water, use of wastewater for irrigation is going to increase even further. But, these unregulated wastewater irrigation practices reveal a range of associated problems that outweigh the benefits and highlights the failures of policies and lack of agricultural extension services. Nevertheless, some Non-Governmental Organisations (NGOs) have taken initiatives to address these issues (see Box 3.3)

Box 3.3 Decentralised wastewater treatment system and usage, Bangalore, India

Gram Swaraj Samithi (GSS), a non-government organization (NGO), partnered with Bremen Overseas Research and Development Association (BORDA) with funding from European Commission and the Federal Ministry for Economic Cooperation and Development, Germany built sanitation complexes fitted with Decentralised Wastewater Treatment System for the population of Ullalu Upanagara, on the outskirts of Bangalore.

It is the first community-based sanitation (CBS) project in India that treated wastewater and harvested rainwater. This made the wastewater suitable for reuse in toilets, bathrooms and for laundry. The facility is operated by the women's Self Help Group (SHG), which is nurtured by GSS. The amount earned from the two units is used for the operation and maintenance of the facility. Surplus amounts are transferred to the SHG's bank account.

At the outset, GSS involved the local government (Panchayat) from the very beginning and initiated education and awareness programmes on critical issues like health, environmental hygiene and sanitation. Community members mobilised contributions from the people of the area for the project. Community needs assessment, revealed the willingness to own and run the sanitation project. BORDA undertook a technical feasibility study. Concurrence from the community on the proposed CBS facility was taken and important stakeholders from the community were sent for a hands-on experience which enhanced the community's practical understanding of such units and their supplementary benefits.

Source http://cbhi-hsprod.nic.in/

References

Australian Bureau of Statistics (2016) Australian environmental-economic accounts 2016, (Report no. 4655.0. ABS). Canberra, Australian Capital Territory, Author

Baez B, Abolafia MY (2002) Bureaucratic entrepreneurship and institutional change: a sense-making approach. J Public Adm Res Theor 12(4):525–552

Bredford A, Brook R, Hunshal C (2003) Wastewater irrigation: Hubli-Dharwad, India. Paper presented at the International Symposium on Water, Poverty and Productive uses of water at the household level, 21–23 Jan 2003, Muldersdrift, South Africa

Buechler SJ, Devi G (2003) Household food security and wastewater-dependant livelihood activities along the Musi River in Andhra Pradesh, India. Report submitted to the World Health Organisation (WHO), Geneva, Switzerland

Cleaver F, Franks T (2005) Water governance and poverty: a framework for analysis. Bradford Centre for International Development (BCID) (Research Paper no. 13). University of Bradford, United Kingdom, Dec 2005

Dillon PJ (2000) Water reuse in Australia: current status, projections and research. In: Dillon PJ (ed) Proceedings of Water Recycling Australia 2000, Adelaide, 19–20 Oct 2000, pp 99–104

Dillon PJ, Toze S, Pavelic P, Ragusa SS, Wright M, Peter P, Martin RR, Gerges NZ, Rinck-Pfeiffer SM (1999) Storing recycled water in an aquifer at Bolivar: benefits and risks. Water 26(5):21–29

Dimitriadis S (2005) Issues encountered in advancing Australia's water recycling schemes. (Research Brief No. 2). Department of Parliamentary Services, Commonwealth of Australia, 2005–06

Dirksen W (2002) Water management structures in Europe. Irrig Drainage 51:199–211

Gaudin Jean-Pierre (1998) Modern governance, yesterday and today: some clarifications to be gained from French government policies. Int Soc Sci J 50(155):47–56

GHD (1978) Planning for the use of reclaimed water in Victoria. Reclaimed Water Committee, Ministry of water resources and water supply, Melbourne, Feb 1978

Gleick P (2000) The changing water paradigm-a look into twenty-first century water resources development. Water Int 25(1):127–138

Grindle MS (2002) Good enough governance: poverty reduction and reform in developing countries. Paper prepared for the Poverty Reduction Group of the World Bank, Nov 2002

Hamilton AJ, Boland Anne-Maree, Stevens D, Kelly J, Radcliffe J, Ziehrl A, Dillon P, Paulin B (2005) Position of the Australian horticultural industry with respect to the use of reclaimed water. Agric Water Manag 71(3):181–209

Hargrave T, Van de Ven AH (2006) Collective action model of institutional innovation. Acad Manag Rev 31(4):864–888

Hobley M, Shields D (2000) The reality of trying to transform structures and processes: Forestry in rural livelihoods, (Working Paper no. 132). Overseas Development Institute (ODI), Feb 2000, London, UK

Huitema D, Meijerink S (2007) Understanding and managing water transitions: a policy science perspective. Paper presented at the Amsterdam Conference on Earth System Governance, Amsterdam, the Netherlands, 24–26 May 2007

Hurlimann A, McKay JM (2006) What attributes of recycled water make it fit for residential purposes? the Mawson Lakes experience. Desalination 187:167–177

International Water Management Institute (2007) IWRM challenges in developing countries: lessons from India and elsewhere. Water Policy Briefing (24) Jan 2007

Joshi D (2004) Secure water-whither poverty? livelihoods in the DRA: a case study of the water supply programme in India. (Research Report). Overseas Development Institute, London

Juwarkar AS, Thawale PR, Jambulkar HP, Juwarkar A (1988) Management of wastewater through crop irrigation–an ecofriendly approach. In: Trivedy RK, Kumar A (eds) Ecotechnology for pollution control and environmental management. Environmental Media, Karad, pp 25–48

Keohane RO (2003) Global governance and democratic accountability. In: Held D, Koenig-Archibugi M (eds) Taming Globalisation: Frontiers of governance. Polity press, Cambridge UK, pp 130–159

Lin JY (1989) An economic theory of institutional change: induced and imposed change. Cato J 9 (1):1–10

Martin RR, Dillon PJ (2002) Aquifer recovery and storage in South Australia. Water 29(2):28–30

McKay JM, Halanaik D (2003) New directions and national leadership in developing water policies in Federations-India and Australia. Paper presented at the ACIAR Conference on institutional issues in water resource allocation: lessons from Australia and implications for India, Beechworth, Australia, 17–18 July 2003

McKay JM (2006) Issues for CEO's of Australian water utilities with the implementation of the integration and ESD requirements in Australian water laws. J Cont Water Res Educ 135 (December):115–130

North DC (1990) Institutions, institutional change and economic performance. Cambridge University Press, Cambridge

Ostrom E (1990) Governing the commons: the evolution of institutions for collective action. Cambridge University press, Cambridge

Radcliffe JC (2004) Water recycling in Australia. A review undertaken by the Australian Academy of Technological Sciences and Engineering, Victoria

Rogers P (2002) Water governance in Latin America and the Caribbean. Inter American Development Bank, Sustainable Development Department, Environment Division, Fortaleza, Brazil

Rogers P, Hall AW (2003) Effective water governance. TEC (Background Paper no. 7). Global Water Partnership-TEC, Sweden

Rotmans J, Kemp R, van Asselt M (2001) More evolution than revolution: transition management in public policy. Foresight 03(01):001–0017

Shah T, van Koppen B (2006) Is India ripe for integrated water resources management? fitting water policy to national development context. Econ Polit Wkly 41(31):3143–3421

Srivastava V (2004) Lessons for India: Australia's water sector reforms. Water and sanitation program-South Asia field note, The World Bank, 55 Lodi Estate, New Delhi, India

Stoker G (1998) Governance as theory: five propositions. Int Soc Sci J 50(155):17–28

Tang SY (1992) Institutions and collective action: self-governance in irrigation. ICS press, San Francisco, CA

Taylor M, Dalton R (2003) Water resources—managing the future. Presented at the Australian Water Association 20th Convention, Perth, Apr 2003

Townsend RE, Pooley SG (1995) Distributed governance in fisheries. In: Hanna S, Munasinghe M (eds) Property rights and the environment-social and ecological issues. The Beijer International Institute of Ecological Economics and the World Bank, Washington, DC

Turrall HN (1998) Hydro-Logic?—Reform in water resources management in developed countries with major agricultural water use: lessons for developing countries. ODI Research Study, Overseas Development Institute, London

United Nations Educational, Scientific and Cultural Organisation (2003) Water for people, water for life. The United Nations World Water Development Report-1. Published jointly by UNESCO, France and Berghahn Books, USA

United Nations Educational, Scientific and Cultural Organization (2006) Water-a shared responsibility. The United Nations World Water Development Report-2. Published jointly by UNESCO, France and Berghahn Books, USA van der Hoek W, Hassan MU, Ensink JHJ, Feenstra S, Raschid-Sally L, Munir S, Aslam R, Ali N, Hussain R, Matsuno Y (2002) Urban wastewater: a valuable resource for agriculture. A case study from Haroonabad, Pakistan. (Research Report-63). International Water Management Institute, Colombo, Sri Lanka

Wegerich K (2001) Institutional change: a theoretical approach (occasional Paper No. 30). Water issue study group, University of London, May 2001

Winpenny J (2003) Financing water for all. Report of the World Panel on Financing Water Infrastructure. Published jointly by World Water Council and Global Water Partnership, March, 2003

Chapter 4
Theoretical Framework

Governance often implies 'good governance', and in achieving god governance there has been a major policy shift in the natural resource management domain in the form of transfer of management responsibilities to users' groups or private utilities. As a result, governance of water resources is now discussed with reference to institutions such as state, community, the market or the individual. At the same time, concepts like 'distributed governance' and 'partnerships' are also becoming popular in the water sector. Partnerships help to pool resources and reduce risks and evidence suggests that people come together when there is a widely acknowledged crisis, a crisis that multiple groups acknowledge to affect their core interests—the concept of collective action. The reasons for collective action vary depending on the circumstances and local conditions.

Irrespective of who owns and manages the resource in question, the role of institutional arrangements or working rules is extremely important. Institutions also shape the incentives for individuals to take certain actions such as cooperating, engaging in collective action, and/or coordinating activities to achieve desired outcomes. Water crises can be best dealt with by cooperation or collaboration between different stakeholders; effective collaboration and rewards (both economic and environmental), can be realised if strong relationships can be developed among the community, the governments, and supporting institutions. The concepts of water governance and institutional change therefore have a number of dimensions, span diverse disciplines, and are related to various theories. Some of the theories and concepts relevant to the present study are discussed below.

4.1 Theory of Institutions

Institutions are part of our daily life and the term 'institutions' in everyday use often refers to ministries, departments, associations, and unions that are actually 'organizations' (Bandaragoda 2000). The two terms 'institutions' and 'organizations' are

© The Author(s) 2017
G. Keremane, *Governance of Urban Wastewater Reuse for Agriculture*,
SpringerBriefs in Water Science and Technology,
DOI 10.1007/978-3-319-55056-5_4

so common in usage that they are often used as synonym. But it should be realized that they have some distinct meanings. Therefore, a clear understanding of these terms is important.

In general sociology, an institution depends on organized, established procedures that are represented as constituent rules of society, or 'rules of the game' (Jepperson 1991, cited in Bandaragoda 2000). We find a similar interpretation by North, who defines institutions as the rules of the game in a society, or the humanly devised constraints that shape human action (North 1991). According to MacDonald and Dyack (2004), institutional arrangements include both administrative arrangements, and the ways that rules regulating water use and reuse are defined. Adopting various definitions of institutions from the literature, Bandaragoda (2000) sums up an institution as:

> a combination of policies; laws, rules and regulations; organizations, their bylaws; operational plans and procedures; incentive and accountability mechanisms; and norms, traditions, and practices (p. 5).

Institutions, therefore, set the ground rules for resource use and establish the incentives, information, and compulsions that guide economic outcomes. They are the sets of the users' rights in relation to the resource in question, and the rules that define what actions they can take in using it. According to Davis and North (1971, pp. 6–7), institutions can be differentiated as those related to the political, legal and social environment of an economic system, and those related to arrangements between economic units that govern the ways in which these units can cooperate and/or compete.

Organisations on the other hand, are groups of individuals with defined roles and bound by some common purpose and rules and procedures to achieve set objectives. Merrey (1996, p. 8) describes organizations as "structures of recognized and accepted roles". North (1990, p. 73) defines organizations as "purposive entities designed by their creators to maximize wealth, income, or other objectives defined by the opportunities afforded by the institutional structure of the society." According to Cernea (1987, cited in Bandaragoda 2000) organizations are networks of behavioural roles arranged into hierarchies to elicit desired individual behaviour and coordinated actions obeying a certain system of rules and procedures; and the hierarchical arrangement is referred to as the 'organizational structure'. Organisations can be government agencies, companies, political parties, churches or non-governmental organisations.

Nevertheless, institutions and organizations are interlinked and this interaction can be perceived in two ways:

Evolution of organizations is influenced by the institutional framework. Eg: In Australia, the creation of NRM Boards or Water Boards followed the articulation of natural resources or water-related policy, and the enactment of water law. Organizations represent a set of norms and behaviours and are in fact institutions. Eg: Water Users Associations (WUAs) in India.

This study focuses on the governance mechanisms, which are comprised of formal and informal institutions and support organisational forms for the production and/or exchange of assets (Bandaragoda 2000; Zenger et al. 2001).

4.1.1 Formal and Informal Institutions

Institutions can be both formal and informal because apart from written laws, rules and procedures, informally established procedures, norms, practices and patterns of behaviour also form part of the institutional framework. This is largely because informally established procedures and norms become 'rules' in their own right, when they are accepted by the society after years of practice. For example, rotational irrigation systems (Ostrom 1990) and contracts, implicit or explicit (Nabli and Nugent 1989, cited in Herath 2002) are institutions because they embody rules and regulations that govern specific activities of the irrigators or the parties involved. However, lack of proper enforcement or disregard of the written laws can make them ineffective, and as a result they replaced by a set of practices referred to as 'rules-in-use' (Bandaragoda 2000). These rules are 'prescriptions that define what actions (or outcomes) are required, prohibited, or permitted, and the sanctions authorized if the rules are not followed' (Ostrom et al. 1994, p. 38). It is 'the final element that structures an action arena' (Schlager and Blomquist 1998, p. 4).

These formal and informal institutions define the behavioural roles of individuals and groups in a given context of human interaction, aiming at a specified set of objectives—like the use of urban wastewater for irrigation in the present case. Zenger et al. (2001, p. 2) distinguish the terms formal and informal institutions as follows:

> Formal institutions are the rules that are readily observable through written documents or rules that are determined and executed through formal position, such as authority or ownership. Formal institutions, thus, include explicit incentives, contractual terms, and firm boundaries as defined by equity positions. Informal institutions, in turn, are the rules based on implicit understandings, being in most part socially derived and therefore not accessible through written documents or necessarily sanctioned through formal position. Thus, informal institutions include social norms, routines, and political processes.

In this study both, the underlying institutions and the organizations as agents of institutional change are considered. As a result a broad interpretation of institutions as suggested by Saleth and Dinar (1999) is used for analysis which covers all three important elements in the institutional framework, namely policies, laws and organizations.

4.1.2 Functions of Institutions

Institutional arrangements or rules-in-use serve as instruments for human cooperation (North 1991) and they can minimize vulnerability, scarcity and conflict; thus enhancing sustainable management of water resources (Marothia 2003). They are applied to resolve conflicts, to maintain a coordinated flow of action and transactions in the society, to indicate what individuals can, must, or may or may not do, and are enforced by collective sanctions (Commons 1931; Marothia 2002, 2003; Herath 2002; Gonce 1971, cited in Marothia 2003). Therefore, institutional arrangements can be subdivided into two sets: (i) operational rules and (ii) collective choice rules (Tang 1992), who points out that operational rules stipulate who can participate as appropriators and providers; what participants may, must, or must not do; and how they will be rewarded and punished, while collective choice rules stipulate the conditions for adopting, enforcing, and modifying the operational rules. Operation rules generally include boundary, allocation, input and penalty rules, which coordinate irrigators in allocation and maintenance activities and collective choice rules interpret the content of institutional arrangements favouring collective action.

Institutions also shape the incentives for individuals to take certain actions such as cooperate, engage in collective action, and/or coordinate activities to achieve desired outcomes. Hence, the incentives individuals have to be involved in group activities also influence the success or failure of the collective action initiatives. This is why some communities succeed while others fail to sustain cooperative behaviour. More on collective action will be discussed later in this chapter. However, in the present context, institutions generally include the operation and maintenance of systems, designing cropping patterns, allocation and scheduling of water, enforcing the rules (or changing them if needed), and regulations governing access to irrigation water by individual farmers (Saleth 1994; Marothia 2003).

Institutions affect individual behaviour and resource management (Kuks 2005). According to Schlager and Blomquist (1998, p. 9),

institutional arrangements are devised to solve shared problems that resource users experience and they are evaluated on the basis of effectiveness – how well the arrangements addressed dilemmas; fairness – how the arrangements addressed distributional issues; ease of monitoring and enforcement – how the arrangements addressed issues of commitment and monitoring; and efficiency.

However, while dealing with Natural Resources Management (water in this case), the focus on institutional regime should be both from a public governance perspective (policy theory) (Kuks 2005; Bressers and Kuks 2005) and a perspective of private property and usage rights (property rights theory) (Ostrom 1990; Bromley 1991). Policy theory concentrates on the effects of resource policies and applied instruments, while property rights theory focuses on bundles of rights and their sustainable management of water resources (Kuks 2005). Further, the institutional arrangements in the context of water resources management are multi-dimensional and there are numerous types of arrangements, characterised by hundreds of different combinations of rules (Ostrom 1990). Nevertheless, these diverse institutional arrangements share at least one thing in common, in the sense that they attempt to address and resolve similar types of issues (Schlager and

Blomquist 1998). Ineffective institutional arrangements lead to a crisis situation which is the case within the water sector.

Crisis situations can be best dealt with by cooperation or collaboration between different stakeholders and by evolving appropriate institutional arrangements or institutions. Likewise, in the context of the present study, institutional and social dimensions cannot be overlooked in the implementation of resource conserving alternative wastewater technologies. The adoption of an alternative technology is dependent directly on the level of acceptance it gains from both the household user and the institutional framework (Frijns and Jansen 1996; Khouri et al. 1994; Veenestra and Alaerts 1996). Further, Frijns and Jansen (1996) have pointed out that although alternative technologies may be less expensive per capita, they often require community efforts and resources from residents. This leads us to the next section(s) which focuses on concepts such as collective action, social capital, partnerships and strategic alliances.

4.2 The Theory of Collective Action

Collective action is mostly discussed in relation to the 'tragedy of the commons' which was made poplar byHardin (1968), through a seminal article. According to him:

> There is the tragedy. Each man in locked into a system that compels him (her) to increase his (her) herd without limit – in a world that is limited. Ruin is the destination toward which all men rush, each pursuing his own best interest in a society that believe in the freedom of the commons (p. 1244).

Overcoming the tragedy of the commons in a real world situation is difficult because much of the world is dependent upon resources that are subject to the possibility of a tragedy of the commons (Ostrom 1992). However, there is a growing consensus among scholars of the commons that collective, community–based efforts hold out the best prospects for efficient management of the Common Pool Resources (Ostrom 1990, 1999, 2000a; Baland and Platteau 1998; Ostrom et al. 1994; Ostrom and Gardner 1993), leading to the emergence of the concept of 'collective action', a response to deal with the tragedy of the commons.

4.2.1 Concept of Collective Action

The term collective action refers to activities that need coordinated efforts by two or more individuals (Meinzen-Dick and Knox 1999; Agarwal and Ostrom 1999; Meinzen-Dick and Di Gregorio 2004; Dantiki 2005). For example, Wade (1979) points out that, in areas where water is problematic for virtually all irrigators, they tend to form a corporate body to deal with common irrigation and cultivation problems.

Collective action is mostly discussed in the context of Common Pool Resource (CPR) management and rightly so, because the literature on the commons is full of instances of collective regulation for natural resource management (Ostrom and Gardner 1993; Schlager and Ostrom 1992; Schlager et al. 1994; Feeny et al. 1990; Lam 1996a, b, 2001; Morrow and Hull 1996; White and Runge 1995; Meinzen-Dick et al. 2000; Agrawal and Gibson 1999; Ostrom et al. 1992). In the event that the state fails to govern the CPRs in an efficient and sustainable way, collective action is seen as an institutional response to the tragedy of the commons.

4.2.2   Reasons for Collective Action

The reasons for collective action vary depending on the circumstances and local conditions. Evidence suggests that people come together when there is a widely acknowledged crisis; one that multiple groups concede is affecting their core interests (UNDP 1999). Individuals may organize due to State failure to govern the CPRs efficiently (Chopra and Gulati 1997). Furthermore, the literature demonstrates that variables such as group size (Baland and Platteau 1998), economic benefits, fairness, trust, and reciprocity (Schmidt et al. 2001) are likely to affect the collective action and cooperation in a given institutional setting. Bardhan (1993) argues that local information can act as an incentive to cooperation or collective action. According to Ostrom (1990, 2000b), a key attribute of collective action is that members invest resources in monitoring and sanctioning the actions of one another, to reduce the probability of free-riding. In addition, the physical and group attributes of the communities influence collective action (Ternstrom 2001; Agarwal 2001; Anand 2003; Mukhopadhyay 2005). Baland and Platteau (1996, cited in Anand 2003, p. 234) provide a comprehensive list of conditions for communities to sustain cooperative behaviour:

> ...user groups must be small, live close to CPRs, and be free to set access and management rules in their own way; the CPRs must be clearly defined and people must have high level of dependence on them; rules as well as techniques of control must be simple and fair; there must be well-established schemes of punishment; costs of monitoring must not be too high; well-known and low cost conflict-resolution mechanisms must be available; crucial decisions must be taken publicly; and some record-keeping and accountability must be provided for

However, mere presence of a crisis is not the only reason for cooperation. Since such action involves multiple groups, separate uncoordinated actions can only lead to missing opportunities to optimise the use of the resource. Therefore, leaders or champions, through their personal motivation, can make partnerships happen (UNDP 1999). Besides, collective-action situations demand participation by all the parties involved.

4.2.2.1 Community Participation and Leadership

Community participation is not new to the water sector, as the governments of several countries, the World Bank and other multinational financing agencies, and donors are promoting the concept of decentralization for managing water resources (World Bank 1993; Mody 2004). According to Blomquist et al. (2005, p. 4), 'decentralization has two components. One is organizing management responsibilities at the river basin scale, which often involves devolution of authority from a central government. The other is involving stakeholders within the basin in decision making and/or operations concerning water resource management activities'.

Participation is a broad term with many variations in its meaning and interpretation; in simple terms it can be expressed as nominal membership, while in a broader sense it can be defined as a process in which people have voice and influence in decision-making (White 1996). In the political sense, it is a principle that allows citizens to take part in the political process (Heyd and Neef 2004). According to Sinha and Suar (2005) there are two dimensions to participation—direct and indirect—while other scholars like Pretty (cited in Heyd and Neef 2004) give a more detailed classification with seven forms of participation. Compiling the information from recent literature on participation Eberlei (2001) argues that participation is/or can be discussed under four and seven stages (see Table 4.1).

In any action that involves participation of multiple actors or stakeholders, leadership plays an important role, however; as Sinha and Suar (2005, p. 127) rightly argue:

> effective leadership can augment collective action by inspiring people, enforcing institutional norms, resolving conflicts, networking with development partners and assuring expected benefits to people.

The literature on leadership offers a great deal of information from different perspectives. Despite their differences, many writers actually emphasize similar points, which McNamara C (1999) contextualises as the 'Leadership Cube(^{TM})', which represents at least 20 different perspectives on leadership and has the

Table 4.1 Stages of participation

Four	Five	Six	Seven
Information-sharing	Information-sharing	Information-sharing	Information-sharing
Consultation	Consultation	Consultation	Consultation
Joint-decision making	Joint-decision making	Joint-decision making	Participation
Initiation and control by stakeholders	Collaboration	Collaboration	Co-determination
	Empowerment	Empowerment	Joint responsibility
		Control by stakeholders	Partnership
			Control by stakeholders

Source Adapted from Eberlei (2001)

Table 4.2 Different dimensions of leadership

Domain	Relevant leadership skills
Leading yourself	Time management, stress management, assertiveness, etc.
Leading other individuals	Coaching, mentoring, delegating, etc.
Leading other groups	Meeting management, facilitation skills, etc.
Leading organizations	Strategic planning, balanced scorecard, etc.
Leading communities	Community organizing, political skills, etc.
Context	Focus of context
Roles	Board Chair, Chief Executive Officer, executive roles, etc.
Traits	Charismatic, influential, ethical, etc.
Orientation	Leadership values
Results-oriented	Timeliness, efficiency, work direction, authority, etc.
Relationship-oriented	Participation, empowerment, relationships, etc.

Source Adapted from McNamara (1999)

following dimensions: (1) five domains of leadership, (2) two contexts of leadership and (3) two orientations of leadership (see Table 4.2).

Irrespective of their different domains, leadership qualities are very important, particularly in the context of leading others—individuals, groups, or communities. These qualities sometime determine the leadership styles, as in case of Indian forest management where four leadership styles—manipulative, authoritarian, participative and charismatic—have been suggested (Sarin 1996, Singh et al. 1996, both cited in Sinha and Suar 2005). However, the perception of which leadership traits would induce greater participation varies from individual to individual. Sinha and Suar (2005) argue that participative style would evoke more participation, while Conger and Kanungo (1987) and Shah (1991) [both cited in Sinha and Suar 2005] are of the opinion that charismatic leaders possess all the abilities to pursue desired goals and that participative style, with the addition of some charisma works better in rural institutions.

4.3   Sustainability and the Theory of Social Capital

In recent times, 'sustainability' and 'sustainable development' has become the catchphrase among politicians, bureaucrats, academics and researchers. Nevertheless, the concept tends to be rather vague and confusing to be used in a wide variety of contexts and without empirical validation (Copus and Crabtree 1996). As defined in the Brundtland report (1987) sustainable development is,

> development that meets the needs of the present without compromising the ability of future generations to meet their own needs.

This implies that, sustainable development is about ensuring a better quality of life for everyone, now and for future generations to come. Although the concept of sustainability has become popular in recent years, it is interpreted differently by specialists in different disciplines. For example, social scientists say a lot about social sustainability; economists deal with economic sustainability and environmentalists deal with environmental sustainability. However, a holistic approach to understand sustainability is to deal with all the three dimensions (Sullivan 2003). However, in this study the focus is more on social sustainability and the role of social capital to attain sustainability.

4.3.1 Social Sustainability

Social sustainability is focused on the development of programs and processes that promote social interaction and cultural enrichment. Social sustainability is related to how we make choices that affect other humans in our global community—'the Earth'. It covers the broadest aspects of business operations and the effect that they have on employees, suppliers, investors, local and global communities and customers. Social sustainability is also related to more basic needs of happiness, safety, freedom, dignity and affection.

Social sustainability emphasizes protecting the vulnerable, respecting social diversity and ensuring that we all put priority on social capital. According to Leviten-Reid (2001),

> For a community to function and be sustainable, the basic needs of its residents must be met. A socially sustainable community must have the ability to maintain and build on its own resources and have the resiliency to prevent and/or address problems in the future.

There are two types or levels of resources in the community that are available to build social sustainability (and, indeed, economic and environmental sustainability)—individual or human capacity, and social or community capacity. Individual or human capacity refers to the attributes and resources that individuals can contribute to their own well-being and to the well-being of the community as a whole. Such resources include education, skills, health, values and leadership. Social or community capacity is defined as the relationships, networks and norms that facilitate collective action taken to improve upon quality of life and to ensure that such improvements are sustainable. To be effective and sustainable, both these individual and community resources need to be developed and used within the context of four guiding principles—equity, social inclusion and interaction, security, and adaptability.

In line with this discussion and the argument that wastewater reuse history is marked with failure of reuse schemes mainly due to lack of community involvement (Po et al. 2004; Hurlimann and McKay 2006) it is important that the policies on wastewater use and management must include the human dimension (Robinson et al. 2005). Therefore, for a reuse scheme to be successful community involvement and/or participation are very important and social infrastructure provides a framework for building shared solutions to a joint problem through negotiation and dialogue processes (Woolcock 2004; Flora and Flora 1993).

4.3.1.1 Community Social Infrastructure—Meaning and Dimensions

Success or failure of a wastewater reuse project largely depends on community participation and involvement. Therefore, it is important to measure the social capital at the community level and maintaining social capital means social sustainability (Keremane and McKay 2007). Putnam (1993) has measured indicators of social capital on provincial and national level; Coleman (1988) has addressed social capital on an individual and household level. This study adopts the concept of 'Social Infrastructure' (Flora and Flora 1993) which is an important mechanism of institutional analysis as a basis for change. According to the authors social infrastructure means that communities begin to look at making slots rather than fitting into slots; entrepreneurial social infrastructure means communities begin to look at risk, both collectively and individually, in a different way (Flora and Flora 1993, p. 58), and it has three major dimensions:

- Symbolic diversity implies a collective or community level orientation toward inclusiveness rather than exclusiveness,
- Resource mobilisation implies that communities must be ever more dependent on their own resources if development is to occur, and
- Quality of linkages or networks means that networks, formal and informal, within the community and with the outside, facilitate the flow of resources, and so broad linkages are important.

The foregoing discussions reveal that 'social capital' is an important factor in achieving social sustainability. So, what is Social Capital and how to measure it?

4.3.2 Social Capital—Meaning and Forms

The role of social capital is vital in policy studies, yet planners and policy makers often fail to understand this concept. There is a growing body of literature that examines the importance of social capital in organising groups to take a collective action and couple social capital with development (Paldam 2000; Carroll and Stanfield 2003; Coleman 1988, 1996; Putnam 2004; Ostrom 1990; Dietz et al. 2003).

In the social capital literature we come across various definitions for social capital. Social capital is a person's or group's sympathy towards another person or group that may produce a potential benefit (Robison et al. 2002). According to Putnam (1993, 2004), social capital includes networks, norms and trust. While discussing social capital we come across many perceptions about the concept; some scholars focus on dense networks (Dietz et al. 2003); some on values and connections (Sen 1995); and some argue that citizen engagement, interpersonal trust, and effective collective action are what form social capital (Rohe 2004). Irrespective of the variations in the perception about social capital, it is apparent that the social capital theory encompasses three aspects: (i) groups and networks, (ii) trust and solidarity, and (iii) cooperation.

Social capital provides a framework for building shared solutions to joint problem through negotiation and dialogue processes that are necessarily and inherently social (Woolcock 2004, p. 188). In innovative and participatory human resource management arrangements, all workers communicate more widely to solve operating problems, thus workers have much richer communication networks, representing higher levels of social capital compared to traditional human resource management systems (Gant et al. 2002). Further, trust is one of the most frequently encountered elements in definitions of social capital and is an indication that social capital plays some important role in sustainable development (Hutchinson 2004; Danchev 2005).

As argued by Danchev (2005, p. 26), 'there is no evidence that deterioration of social capital can be compensated for by the rise of other forms of capital; on the contrary, when we observe the worsening of social capital, all other forms of capital including development deteriorate'. Therefore, building social capital can be a powerful mechanism for planners who seek to promote greater equity in and across cities (Vidal 2004 p. 164). Thus, social capital makes a difference in terms of a community's ability to solve its own problem (Flora 1995).

4.4 Water Governance—Public Versus Distributed Governance

Traditionally, water has been managed and controlled by many societal entities, with substantial government involvement. Accordingly, Kuks (2005), discussing water governance, regards it as a collective action with respect to water issues that is not restricted to government action, but includes the involvement and participation of non-public stakeholders. Water governance deals with extracting, distributing, and using water within current institutions, and must address the complexity of the institutional context in which collective action is being pursued. Since this study aims to analyse waste-water governance structures in Australia and India, an attempt will be made in this section to identify the dimensions which relate to them.

The governance of water is generally linked to public governance since the common perception is that public agencies possess all the required resources, expertise, and authority to manage water resources. However, the situation is different today and it is not enough to understand a policy sector (water policy in this case) in terms only of policy goals and instruments. This is because public authorities and target groups, and consequently their actions, are influenced by their administrative capacity for policy implementation, different perceptions of the problems at stake, the positions and linkages of the actors in the policy network, and the relations between different stakeholders (Bressers and Kuks 2005), who further argue that governance structure can be analysed along five dimensions:

- levels and scales of governance;
- actors in the policy network;
- problem perceptions and policy objectives;
- strategy and instruments; and
- responsibilities and resources for implementation.

4.4.1 Distributed Governance

The shift in water resources management paradigm has changed the general perception about governing water resources and services. Experiences around the world clearly indicate that, acting alone, neither the public nor the private sector can meet the continually growing demand for water, waste, and energy services (UNDP 1999). New approaches that involve collaboration among an increasing number of stakeholders are urgently needed and hence, governance of water resources is now discussed with reference to different institutions such as state, community, market or individual (Pradhan 2000; Marothia 2002). Therefore, compared to the traditional water governance system, the general perception on governing water resources and services has changed over time. It is now believed that water governance is more effective with broader participation by civil society and private enterprise, and consequently concepts such as distributed governance and partnerships are gaining popularity among water managers and policy makers. Nevertheless, there is always an ongoing debate between the policy makers, planners, and researchers involved about the pros and cons of partnerships and not all partnerships have been a success (Grimsey and Lewis 2004). Like any other sector, a number of strategic alliances between the private and public sectors to provide improved delivery services have been seen in the water sector. It becomes imperative to understand strategic alliances and partnerships.

4.4.1.1 Understanding Strategic Alliances

Over the past 25 years, collaborative activities have become more prominent and extensive in all sectors in many nations. As a result, we have witnessed a surge in strategic alliances among competing firms or companies located in the same country or across national boundaries (Murray 1995). This rapid growth since the 1980s is viewed as further evidence of globalisation (Narula and Hagedoorn 1999).

Strategic alliances often represent a variety of collaborative agreements among competing firms that are more than standard customer-supplier relationships or venture capital investments in nature (Terpstra and Simonin 1993). One type of collaborative engagement often observed at a domestic level is partnerships between business, government, and civil society to address social issues and causes (Selsky and Parker 2005). These authors further point out that such partnership are formed to address challenges such as economic development, education, health care, poverty alleviation, community capacity building, and environmental sustainability.

In his seminal paper entitled 'Symbiotic Marketing', Alder first recognized the possibilities of forming strategic alliances (Murray 1995); since then, there has been a rapid growth in domestic and international alliances.

Parkhe (1993) perceives strategic alliances as innovative and interesting forms of relationships between organizations, which differ from the traditional interactions of organizations. Wheelen and Hungar (2000, p. 125) suggest that a strategic alliance is 'an agreement between firms to do business together in ways that goes beyond normal company-to-company dealings, but fall short of merger'. Narula and Hagedoorn (1999), argue that the terms strategic alliance, collaborative agreement, and network are often used as synonyms. They further specifically define strategic alliances as, 'inter-firm cooperative agreements which are intended to affect the long-term product market positioning of at least one partner' (Narula and Hagedoorn 1999, p. 284).

In summary, as stated by Harbison and Pekar (1998), a strategic alliance is defined as a cooperative arrangement between two or more companies where:

- A common strategy is developed in conformity and all parties adopt a win-win attitude.
- The relationship is reciprocal, with each partner prepared to share specific strengths with each other, thus lending power to the enterprise.
- Pooling of resources, investment, and risks occurs for mutual (rather than individual) gain.

In general, these alliances are intended to allow the parties involved to attain common goals in a more efficient and timely manner than if they were acting alone and, in some cases, to attain goals that they would not be able to achieve using only their own resources. A review of current relevant literature and empirical studies indicates the most common reasons for forming strategic alliances. Some of these are mentioned below.

Alliances are formed to obtain technology, gain access to specific markets, reduce financial and political risks, and achieve competitive advantage (Wheelen and Hungar 2000). According to Likhi and Sushil (2005), strategic alliances are formed for a variety of reasons, which include entering new markets, reducing manufacturing costs, developing and diffusing new technologies, accelerating product introduction, and overcoming legal and trade barriers. Kanter (1994) suggests that organizations create alliances in their quest to compete against fast and agile competitors. Rai et al. (1996) are of the opinion that strategic alliances provide an effective means to improve on both the economies of scale and scope offered by traditional modes of organization.

The above discussion makes it clear that companies may form alliances in order to gain access to the management strengths or regulatory expertise of another company. Alliances may provide sources of raw materials and can be a means to overcome legal and trade barriers. In some cases, a company with a product may form an alliance with another company that has an established distribution system that the first company cannot create for itself without incurring great cost and delays in market penetration.

However, strategic alliances do not always achieve their desired results. As pointed out by Hamel et al. (1989), uncertainty about the behaviour of partners can be a cause for concern, leading to unstable and conflicting relationships. Parkhe (1993) attributes the failure of strategic alliances to a significant dearth of theoretical and empirical research on the topic.

In the context of this study, the focus is on strategic alliances at the domestic level, among Private, Public, and the community organisations—the three main societal sectors. One such type is public–private partnerships, meaning working arrangements based on a mutual commitment between a public sector organization and an organization outside the public sector (Bovaird 2004).

4.5   Public–Private Partnerships (PPP) in the Water Sector

Generally, public sector means the 'Government' and any other entity that is non-governmental is the private sector. However, it is difficult to say precisely and thoroughly what we mean by 'public' and 'private' because in literature we find a number of methods or approaches to handle the issue namely common sense approach, practical definitions, analytical definitions and denotative approaches (see Rainey et al. 1976 for detailed discussion about these methods). According to Lachman (1985, p. 671),

> profit making business firms commonly represent the private sector, and nonprofit service or government regulatory agencies commonly represent the public sector.

Within the water sector, public sector means the 'Government' while, private sector may include private businesses, non-governmental organizations (NGOs) and community-based organizations (CBOs). Some researchers argue that, ownership and management by user cooperatives or the community is also included under private sector (Turral 1995). However, this study sees the water users co-operatives or water users associations (WUAs) as a third category—'self governed organisations' (Ostrom 1992). But, again these self-governed organisations (for example the WUAs in India) may be categorised into two groups: (1) those which own the irrigation system and are fully responsible to control and manage the system, and (2) those which have partial control over the control and management of irrigation systems with the government agencies owning the system. The focus of this study is on the second category which includes systems that have been turned over by the government to user groups for management. This category is also known by other terms such as 'irrigation communities', 'communal irrigation', and 'farmer-managed irrigation systems" and there is considerable variety in the size, technology, and organization of these self-governing systems (Merrey 1996).

According to Grimsey and Lewis (2004, p. 2), 'any relationship involving some combination of the private, and public sectors is prone to be labelled a partnership'. Sharing of responsibility and/or authority between the parties involved is an essential ingredient of partnership (Townsend and Pooley 1995). According to Caplan et al. (2001), a partnership is just a means for delivering the project objectives; therefore, the need today is to implement and enforce the rules under which private or public agencies are made efficient and responsive to social needs and desires (Wolff and Palaniappan 2004). Worldwide, numbers of examples of this cooperation or collaboration in various forms exist (see Grimsey and Lewis 2004); one of the most promising forms of partnership is the Public–Private Partnership (UNDP 1999). In the water sector, the pressing need for more investment in water infrastructure, coupled with constrained government resources, is the main reason behind the emergence of Public–Private Partnerships.

Public–Private Partnerships, popularly known as PPPs, describe a spectrum of possible relationships between public and private actors for the cooperative provision of infrastructure services (UNDP 1999). Grimsey and Lewis (2004, p. 2) see PPPs as a "contract for a private entity to deliver public infrastructure-based service". In line with this definition, in the context of the water sector, PPPs refer to "public entity entering into a contractual agreement with private sector to take over some or all of its activities related to water management" (UNDP 1999; OECD 2003; ADB 2000). Thus, through PPPs, the social responsibility, environmental awareness, and local knowledge of the public sector can be combined with the innovation, access to finance, technology, managerial efficiency, and entrepreneurial spirit of the private sector in order to solve urban problems (UNDP 1999).

Generally, PPPs are misunderstood as 'privatisation', but they differ from privatisation. Grimsey and Lewis (2004) state that two major differences—regulation through contract and the lack of government disengagement in case of the PPPs—differentiate them from privatisation. In privatisation, the management and ownership of the water infrastructure are completely transferred to the private sector, while, in the case of a PPP, the ownership of the assets of the water utility remains with the government, and only the management is contracted out to professional management, which is held accountable and has appropriate incentives to ensure effective delivery and reduce waste (OECD 2003; ADB 2000). The only essential criterion with respect to PPP is some degree of private participation in the delivery of traditionally public-domain services. However, as stated by Grimsey and Lewis (2004, p. 55), "PPPs might still be seen as privatisation in all but name, as they are by many public sector unions".

Table 4.3 Options for private sector involvement in water sector and allocation of responsibilities

Responsibilities	Formal models of PSP in water sector				
	Service contract	Management contract	Lease/affermage	Concession	Divestiture
Asset ownership	Public	Public	Public	Public	Private
Capital investment	Public	Public	Public	Private	Private
Commercial risk	Public	Public	Shared	Private	Private
Operations/maintenance	Private/public	Private	Private	Private	Private
Contract duration	1–2 years	3–5 years	8–15 years	20–30 years	Indefinite
Description[a]	Short-term agreements for a specific task	Government transfers certain O&M responsibilities but retains other	Government transfers all O&M responsibilities	Private agency manages the entire utility and government retains the ownership of assets	Government transfer the water business to private agency including infrastructure on a permanent basis

Source Modified from Budds and McGranahan (2003)
Note [a]See Budds and McGranahan (2003) for further description of the model

4.5.1 Options for Public–Private Partnerships

Options for PPPs can be tailored to satisfy very specific needs, however, for PPP to work to the advantage of the concerned country, it is always important to ensure that social and environmental issues are taken into account (Requena and Lamrani 2002). The literature provides us with a wide range of options for involving the private sector that might be applicable to the water (irrigation) sector (OECD 2003; UNDP 1999; Finlayson 2002; Requena and Lamrani 2002). Table 4.3 illustrates the different forms of PPP and the allocation of public/private responsibilities across these forms.

According to Pierson and McBride (1996), cited in Grimsey and Lewis (2004, p. 2), the mechanics of the arrangements can take many forms and may incorporate some or all of the following features:

- the public sector entity transfers land, property or facilities controlled by it to the private sector entity (with or without payment in return) for the term of the arrangement;
- the private sector entity builds, extends or renovates a facility;
- the public sector entity specifies the operating services of the facility;
- services are provided by the private sector entity using the facility for a defined period of time (usually with restrictions on operations standards and pricing); and
- the private sector entity agrees to transfer the facility to the public sector (with or without payment) at the end of the arrangement.

References

Abu Madi M, Braadbaart O, Al-Sa'ed R, Alaerts G (2003) Willingness of farmers to pay for reclaimed wastewater in Jordan and Tunisia. Water Sci Technol Water Supply 3(4):115–122

Abu-Madi M (2004) Incentive system for wastewater treatment and reuse in irrigated agriculture in the MENA region: evidence from Jordan and Tunisia. Unpublished Ph.D. dissertation submitted to the Delft University of Technology, Delft

Agarwal A (2001) Common property institutions and sustainable governance of resources. World Dev 29(10):1649–1672

Agarwal A, Ostrom E (1999) Collective action, property rights, and devolution of forest and protected area management. Paper presented at Workshop on collective action, property rights and devolution of natural resources management, Puerto Azul, The Philippines, 21–25 June 1999

Agrawal A, Gibson CC (1999) Enchantment and disenchantment: the role of community in natural resource conservation. World Dev 27(4):629–649

Anand PB (2003) From conflict to co-operation: some design issues for local collective action institutions in cities. J Int Dev 15:231–243

Asano T (2001) Water from (waste) water—the dependable water resource. Paper presented at the 11th Stockholm Water Symposium, Stockholm, Sweden, 12–18 Aug 2001

Asian Development Bank (2000) Public–Private Partnerships in the Social Sector: Issues and Country Experiences in Asia and the Pacific, (ADBI Policy Paper No. 1). Tokyo, Japan: Author

Baland J, Platteau J (1998) Division of the commons: a partial assessment of the new institutional economics of land rights. Am J Agric Econ 80(3):644–650

Bandaragoda DJ (2000) A framework for institutional analysis for water resources management in a river basin context, (working paper 5). International Water Management Institute. Colombo, Sri Lanka

Bardhan P (1993) Symposium on management of local commons. J Econ Perspect 7(4):87–92

Blomquist W, Dinar A, Kemper K (2005) Comparison of institutional arrangements for river basin management in eight basins. World Bank policy research working paper 3636, June 2005

Blomquist W, Heikkila T, Schlager E (2004) Building the agenda for institutional research in water resource management. J Am Water Resour Assoc 40(4):925–936

Bovaird T (2004) Public–private partnerships: from contested concepts to prevalent practice. Int Rev Admin Sci 70(2):199–215

Bressers H, Kuks SMM (2005) Integrated regimes and sustainable use of natural resources: a multiple case study analysis. presented at the conference on sustainable water management: comparing perspectives from Australia, Europe and the United States, 15–16 Sept 2005, Canberra, Australia

Bromley WD (1991) Environment and economy: property rights and public policy. Blackwell, Cambridge

Brundtland Report (1987) World Commission on Environment and Development. Our common future. United Nations

Budds J, McGranahan G (2003) Are the debates on water privatization missing the point? Experiences from Africa, Asia and Latin America. Environ Urbanization 15:87–113

Caplan K, Heap S, Nicol A, Plummer J, Simpson S, Weiser J (2001) Flexibility by design: lessons from multi-sector partnerships in water and sanitation projects, London. Business Partners for Development, Water and Sanitation Cluster. Retrieved from http://www.bpdws.org/english/docs/flexibility.pdf

Carroll CM, Stanfield JR (2003) Social capital, Karl Polanyi, and American social and institutional economics. J Econ Issues 37(2):397–404

Chopra K, Gulati SC (1997) Environmental degradation and population movements: the role of property rights. Environ Resour Econ 9:383–408

Coleman SJ (1996) The possibility of a social welfare function. Am Econ Rev 56:1105–1122

Coleman SJ (1988) Social capital in the creation of human capital. Am J Sociol 94:95–120

Commons JR (1931) Institutional economics. Am Econ Rev 21:649–657

Cooper M (2006) Social sustainability in Vancouver. Research paper commissioned from Canadian Policy Research Networks, Ottawa, Ontario

Copus AK, Crabtree JR (1996) Indicators of socio-economic sustainability: an application to remote rural Scotland. J Rural Stud 12(1):41–54

Danchev A (2005) Social capital influence on sustainability of development (case study of Bulgaria). Sustain Dev 13(2005):25–37

Dantiki S (2005) Organizing for peace: collective action problems and humanitarian intervention. J Mil Strateg Stud 7(3):1–21

Davis LE, North DC (1971) Institutional change and American economic growth. Cambridge University Press, Cambridge

Dietz T, Ostrom E, Stern P (2003) The struggle to govern the commons. Science 302:1907–1912

Eberlei W (2001) Institutionalised participation in processes beyond the PRSP. Study commissioned by GTZ, Eschborn, Institute for Development and Peace (INEF), Gerhard-Mercator-University, Duisburg

Feeny D, Berkes F, McCay JB, Acheson MJ (1990) The tragedy of the commons: twenty-two years later. Hum Ecol 18(1):1–19

Finlayson J (2002) The what and why of public–private partnerships. Policy Perspect 9(1):1–6

Flora CB (1995) Social capital and sustainability: agriculture and communities in the Great Plains and Corn Belt. Res Rural Sociol Dev 6:227–246

Flora CB, Flora JL (1993) Entrepreneurial social infrastructure: a necessary ingredient. Ann Acad Soc Polit Sci 529(September):48–58

Frijns J, Jansen M (1996) Institutional requirements for appropriate wastewater treatment systems. Paper presented at the workshop on sustainable municipal waste water treatment systems, Leusdan, The Netherlands, 12–14 Nov 1996

Gant J, Ichniowksi C, Shaw K (2002) Social capital and organizational change in high-involvement and traditional work organizations. J Econ Manage Strategy 11(2):289–328

Grimsey D, Lewis MK (2004) Public private partnerships: the worldwide revolution in infrastructure provision and project finance. Edward Elgar Publishing Limited, UK and USA

Hamel G, Doz Y, Prahalad CK (1989) Collaborate with your competitors—and win. Harvard Bus Rev 67(January–February):133–139

Harbison JR, Pekar P Jr. (1998) Smart alliances. Jossey-Bass Publishers, San Francisco

Hardin G (1968) The tragedy of the commons. Science 162:1243–1248

Herath G (2002) Issues in irrigation and water management in developing countries with special reference to institutions. In: Donna Brennan (Ed) Water policy reform: lessons from Asia and Australia. The Australian Centre for International Agricultural Research (ACIAR) proceedings no. 106, Canberra, Australia

Heyd H, Neef A (2004) Participation of local people in water management-evidence from the Mae Sa watershed, Northern Thailand, (discussion paper No. 128). Environment and production technology division, International Food Policy Research Institute (IFPRI), Washington, DC

Hutchinson J (2004) Social capital and community building in the inner city. J Am Plann Assoc 70 (2):168–175

Hurlimann A, McKay JM (2006) What attributes of recycled water make it fit for residential purposes? The Mawson Lakes experience. Desalination 187:167–177

Kanter RM (1994) Collaborative advantage: the art of alliances. Harvard Bus Rev 72(4):142–149

Keremane GB, McKay J (2007) Successful wastewater reuse scheme and sustainable development: a case study in Adelaide. Water Environ J 21(2):83–91

Khouri N, Kalbermatten JH, Bartone CR (1994) The reuse of wastewater in agriculture: a guide for planners. (Water and sanitation report No. 6). UNDP-World Bank water and sanitation program, the World Bank, Washington, D.C

Kuks, S. M. M. (2005). The Evolution of national water regimes in Europe: Transitions in water rights and water policies. Paper presented at the conference on sustainable water management: comparing perspectives from Australia, Europe and the United States, Canberra, Australia, 15–16 Sept 2005

Lachman R (1985) Public and private sector differences: CEOs' perceptions of their role environments. Acad Manag J 28(3):671–680

Lam WF (1996a) Improving the performance of small-scale irrigation systems: the effects of technological investments and governance structure on irrigation performance in Nepal. World Dev 24(8):1301–1315

Lam WF (1996b) Institutional design of public agencies and coproduction: a study of irrigation associations in Taiwan. World Dev 24(6):1039–1054

Lam WF (2001) Coping with change: a study of local irrigation institutions in Taiwan. World Dev 29(9):1569–1592

Leviten-Reid E (2001) Opportunities 2000: multisectoral collaboration for poverty reduction. Final evaluation report. Caledon Institute of Social Policy, Ottawa

Likhi D, Sushil (2005) The importance of situation, actors and process in management of strategic alliances. Glob Busi Rev 6(1):29–39

MacDonald DH, Dyack B (2004) Exploring the institutional impediments to conservation and water reuse—National issues. CSIRO Land and Water Client Report

Marks RF (1993) Appropriate sanitation options for Southern Africa. Water Sci Technol 27(1):1–10

Marothia DK (2002) Institutional arrangements for participatory irrigation management: Initial feedback from Central India. In: Brennan D (Ed) Water policy reform: lessons from Asia and Australia, The Australian Centre for International Agricultural Research (ACIAR) proceedings no. 106, Canberra, Australia

McNamara C (1999) Leadership Cube(TM) to contextualise leadership. Authenticity Consulting LLC, (Copyright 1997–2007). Retrieved from http://www.managementhelp.org/ldrship/carters.htm

Marothia DK (2003) Enhancing sustainable management of water resources in agricultural sector: the role of institutions. Indian J Agric Econ 58(3):406–427

Meinzen-Dick R, Knox A (1999) Collective action, property rights, and devolution of natural resource management: a conceptual framework. Paper presented at the Workshop on collective action, property rights and devolution of natural resources management, Puerto Azul, The Philippines, 21–25 June 1999

Meinzen-Dick R, Di Gregorio M (2004). Collective action and property rights for sustainable development. Focus 11, CAPRi: CGIAR system wide program on collective action and property rights

Meinzen-Dick R, Raju KV. Gulati A (2000) What affects organization and collective action for managing Resources? Evidence from Canal Irrigation Systems in India (EPTD discussion paper No. 61). Washington DC IFPRI

Merrey DJ (1996) Institutional design principles for accountability in large irrigation systems, (research report 8). International Irrigation Management Institute (IIMI), Colombo, Sri Lanka

Mills R, Asano T (1996) A retrospective assessment of water reclamation projects. Water Sci Technol 33(10–11):59–70

Mody J (2004) Achieving accountability through decentralization: lessons for integrated river basin management. Policy research working paper 3346. The World Bank, Washington, DC

Morrow EC, Hull RW (1996) Donor-Initiated common pool resource institutions: the case of the Yanehsa forestry cooperative. World Dev 24(10):1641–1657

Mukhopadhyay P (2005) Now that your land is my land...does it matter? A case study in Western India. Environ Dev Econ 10:87–96

Murray JY (1995) Patterns in domestic vs. international strategic alliances: an investigation of US multinational firms. Multinational Busi Rev 3(2):7–16

Narula R, Hagedoorn J (1999) Innovating through strategic alliances: moving towards international partnerships and contractual agreements. Technovation 19(5):283–294

North DC (1990) Institutions, institutional change and economic performance. Cambridge University Press, Cambridge

North DC (1991) Institutions. J Econ Perspect 5(1):97–112

Organization for Economic Co-operation and Development (2003) Public–private partnerships in the Urban Water Sector. OECD Policy Brief

Oron G, Campos C, Gillerman L, Salgot M (1999) Wastewater treatment and reuse for agricultural irrigation in small communities. Agric Water Manag 38:223–234

Ostrom E (1990) Governing the commons: the evolution of institutions for collective action. Cambridge University Press, Cambridge

Ostrom E (1992) Crafting institutions for self-governing irrigation systems. ICS press, San Francisco

Ostrom E (1999) Coping with the tragedies of the commons. Annu Rev Polit Sci 2:493–535

Ostrom E (2000a) Reformulating the commons. Swiss Polit Sci Rev 6(1):29–52

Ostrom E (2000b) Collective action and the evolution of social norms. J Econ Perspect 14(3):137–158

Ostrom E, Gardner R (1993) Coping with asymmetries in the commons: self-governing irrigation systems can work. J Econ Perspect 7(4):93–112

Ostrom E, Walker J, Gardner R (1992) Covenants with and without a sword: self-governance is possible. Am Polit Sci Rev 86(2):404–417

Ostrom E, Gardner R, Walker J (1994) Rules, games, and common-pool resources. University of Michigan Press, Ann Arbor

Paldam M (2000) Social capital: one or many? Definition and measurement. J Econ Surv 14 (5):629–653

Parkhe A (1993) Strategic alliance structuring: a game theoretic and transaction cost examination of inter-firm cooperation. Acad Manag J 36(4):794–829

Po M, Juliane K, Nancarrow BE (2004) Literature review of factors influencing public perceptions of water reuse. Australian water conservation and reuse research program. CSIRO

Pollnac RB (1988) Evaluating the potential of fishermen's organizations in developing countries. International Center for Marine Resource Development, University of Rhode Island, Kingston

Pollnac RB, Crawford BR (2000) Discovering factors that influence the success of community based marine protected areas in the Visayas, Philippines. (Coastal management report No. 2229). PCAMRD Book Series No. 33. Coastal Resources Centre, University of Rhode Island, Narrangansett, RI, USA, and Philippine Council for Aquatic and Marine and Development, Los Banos, Philippines

Pradhan P (2000) Farmer managed irrigation systems in Nepal at the crossroad. Paper presented at the 8th Biennial Conference of the International Association for the Study of Common Property (IASCP), Bloomington, Indiana, May 30–4 July 2000

Putnam DR (1993) The prosperous community: social capital and public life. Am Prospect 13:35–42

Putnam DR (2004) Preface. J Am Plan Assoc 70(2):141–143

Rai A, Borah S, Ramaprasad A (1996). Critical success factors for strategic alliances in the information technology industry. Decision Sciences, pp 141–155

Rainey HG, Backoff RW, Levine CH (1976) Comparing public and private organizations. Publ Adm Rev 36(2):233–244

Requena S, Lamrani H (2002) Options and challenges for PPP in irrigation. Paper presented at the Middle East and North Africa Regional Consultation for the 3rd World Water Forum, Spain, 10–12 June 2002

Robinson KG, Robinson CH, Hawkins SA (2005) Assessment of public perception regarding wastewater reuse. Water Sci Technol Water Supply 5(1):59–65

Robison JL, Schmid AA, Siles EM (2002) Is social capital really capital. Rev Soc Econ 60(1):1–21

Rohe MW (2004) Building social capital through community development. J Am Plan Assoc 70 (2):158–164

Saleth RM (1994) Groundwater markets in India: a legal and institutional perspective. Indian Econ Rev 29(2):157–176

Saleth RM, Dinar A (1999) Evaluating water institutions and water sector performance, (technical paper No. 447). World Bank, Washington DC

Schlager E, Blomquist W (1998) Resolving common pool resource dilemmas and heterogeneities among resource users. Paper presented at 'Crossing Boundaries'-Seventh annual conference of the international association for the study of common property, Vancouver, Canada, 10–14 June 1998

Schlager E, Ostrom E (1992) Property-rights regimes and natural resources: a conceptual analysis. Land Econ 68(3):249–262

Schlager E, Blomquist W, Tang YS (1994) Mobile flows, storage, and self-organized institutions for governing common pool resources. Land Econ 70(3):294–317

Schmidt D, Shupp R, Walker J, Ahn TK, Ostrom E (2001) Dilemma games: game parameters and matching protocols. J Econ Behav Organ 46:357–377

Selsky JW, Parker B (2005) Cross-sector partnerships to address social issues: challenges to theory and practice. J Manag 31(6):849–873

Sen A (1995) Rationality and social choice. Am Econ Rev 85(1):1–24

Sinha H, Suar D (2005) Leadership and people's participation in community forestry. Int J Rural Manag 1(1):125–143

Standards Australia International (2003). Australian Standard corporate social responsibility (AS 8003-2003), Standards Australia International, Sydney, NSW, 23 June 2003

Sullivan P (2003) Fundamentals of Sustainable agriculture: applying the principles of sustainable farming. National Centre for Appropriate Technology, Fayetteville, Arkansas

Tang SY (1992) Institutions and collective action: self-governance in irrigation. ICS Press, San Francisco, CA

Ternstrom I (2001) Cooperation or conflict in common pools. SSE/EFI working paper series in Economics and Finance No. 428

Terpstra V, Simonin BL (1993) Strategic alliances in the triad: an exploratory study. J Int Mark 1 (1):4–25

Townsend RE, Pooley SG (1995) Distributed governance in fisheries. In: Hanna S, Munasinghe M (eds) Property rights and the environment-social and ecological issues. The Beijer International Institute of Ecological Economics and the World Bank, Washington, D.C

Turral HN (1995) Recent trends in irrigation management changing directions for the public sector, natural resource perspectives, number 5, September 1995. Overseas Development Institute, London

United Nations Development Programme (1999) Public–private partnerships for the urban environment. UNDP/PPPUE conference paper series, volume III, Published jointly by UNDP, New York and Carl Duisberg Gesellschaft e.V.(CDG), Germany

Veenestra S, Alaerts G (1996) Technology selection for pollution control. In: Balkema A, Aalbers H, Heijndermans E (eds) Workshop on sustainable municipal waste water treatment systems, Leusdan, The Netherlands, 12–14 Nov 1996, pp 17–40

Vidal CA (2004) Building social capital to promote community equity. J Am Plan Assoc 70 (2):164–168

Wade HR (1979) The social response to irrigation: an indian case study. J Dev Stud 16(1):3–26

Wheelen TL, Hungar DJ (2000) Strategic management and business policy, 7th edn. Addison-Wesley, New York, pp 125–134

White S (1996) Depoliticizing development: the uses and abuses of participation. Dev Pract 6 (1):6–15

White AT, Runge FC (1995) The Emergence and evolution of collective action: lessons from watershed management in Haiti. World Dev 23(10):1683–1698

Wolff PE, Palaniappan M (2004) Public or private water management? Cutting the Gordian Knot. J Water Resou Plan Manage 130(1):1–3

Woolcock M (2004) Why and how planners should take social capital seriously. J Am Plan Assoc 70(2):183–189

World Bank. (1993). Water resources management. A World Bank policy paper. Washington, D.C

Zenger TR, Lazzarini SE, Poppo L (2001) Informal and formal organisation in New Institutional Economics. Retrieved from http://papers.ssrn.com/sol3/papers.cfm?abstract_id=319300

Chapter 5
Study Design and Methods

This study evaluated water laws, policies, and institutions related to urban wastewater reuse in South Australia, which is largely formal, innovative, and involves some form of bureaucratic entrepreneurship. The study also included a study in India, typified by its informal and unregulated use of wastewater for irrigation, thereby allowing comparison of formal (regulated) and informal (unregulated) use of urban wastewater for agricultural irrigation.

5.1 Study Design

This study adopted a case study research strategy which is one of several ways of doing social science research. Case studies illustrate the general trends of the events leading to success or failure of an effort. Case study evaluations can cover both process and outcomes; they can include both quantitative and qualitative data (Tellis 1997). According to Yin (1994):

> case studies are the preferred strategy when "how" or "why" questions are being posed, when the investigator has little control over events, and when the focus is on a contemporary phenomenon within some real-life context' (p. 1).

Case studies can be exploratory, descriptive or explanatory and use of each of these strategies depends on three conditions: (i) the type of research question posed, (ii) the extent of control an investigator has over actual behavioural events, and (iii) the degree of focus on contemporary as opposed to historical events (Yin 1994, p. 4). In the present case, the focus is on finding how the community has managed the (re)use of urban wastewater for agricultural irrigation by means of formal or informal arrangements. Therefore, following Yin's (1994) argument, a case study approach can cope better with the technically distinctive situation, can rely on multiple sources of evidence, and can benefit from the prior development of theoretical propositions to guide data collection and analysis. Furthermore, case study

© The Author(s) 2017
G. Keremane, *Governance of Urban Wastewater Reuse for Agriculture*,
SpringerBriefs in Water Science and Technology,
DOI 10.1007/978-3-319-55056-5_5

research can include single and multiple case studies. In this instance, since the study includes three wastewater irrigation schemes (2 in Australia and 1 in India), the situation is that same investigation will include multiple case studies.

5.1.1   Sampling Design

The major method of sampling employed in this study is purposive sampling. In addition, the snow-ball sampling method, a sub-set of purposive sampling, is used to select sample households for interview survey (particularly in the case of the Virginia pipeline scheme in South Australia). Purposive sampling is a form of non-probability sampling (Polit and Hungler 1999, p. 284). In purposive sampling, we sample with a purpose in mind. According to Trochim (2001), purposive sampling can be very useful for situations where the need is to reach a targeted sample quickly, and where sampling for proportionality is not the primary concern. With a purposive sample, you are likely to get the opinions of your target population, but you are also likely to overweight those subgroups in your population that are more readily accessible.

5.1.1.1   Selection of the Schemes

The universe of inquiry for this study comprises two reclaimed water irrigation schemes in South Australia, with their communities, and a third case study, in Andhra Pradesh, India. In the case of the Australian case studies, an initial exploration of the study area was conducted well before the surveys were initiated. The objective of this exercise was mainly to familiarize with the study area and the schemes.

Unlike the case studies in Australia, wastewater use in India is indirect, which means that wastewater is disposed of in rivers, and the contaminated river water is used for irrigation (van der Hoek et al. 2002). Most of this reuse occurs along the many Indian peninsular rivers for agricultural irrigation, and the Musi River, flowing across Andhra Pradesh, is one of these (Buechler and Devi 2003). This was selected as the third case study mainly as it provides an opportunity to compare formal and informal wastewater reuse for agricultural irrigation and partly to fulfill the project requirements of Australian Centre for International Agricultural Research (ACIAR), which parlty funded this study.

5.1.1.2   Selection of Respondents

Apart from their geographic locations, the irrigation schemes under study differed in the composition of the irrigators receiving water from the schemes. Therefore, selection of respondents varied between these schemes; the actual processes adopted in each case are explained below.

Virginia Pipeline Scheme

The selection of sample population for household interviews for the Virginia pipeline scheme was carried out through snowball sampling, one of the strategies for data gathering within the purposive sampling frame. The initial visits to the study sites and discussions held with the concerned authorities had revealed that, due to the privacy policy and confidentiality issues, it was difficult to obtain a list of the irrigators using water from the scheme. Given that situation, snowball sampling appeared to be the most suitable technique. Snowball sampling is an approach for locating informants in the case of certain hard-to-reach subgroups of the population (Patton 1990). The basis here is that an initial contact from the hard-to-reach subgroup may then introduce the researcher to a network of further informants. Using this approach, a few potential respondents are contacted and asked whether they know of anybody with the characteristics that the researcher is looking for in his/her research.

Since the water users on the Virginia pipeline scheme constituted people from different ethnic communities it was difficult to get the exact composition of irrigators associated with the scheme due to privacy policies. So it was decided to obtain the overall composition of the population in the region. According to the Playford City Council Community Profile, the total population in Virginia Township consists of people from Vietnam, Cambodia, Greece, Italy, Serbia, Turkey and Macedonia. However, for this study the total population was broadly classified into two groups: (1) English-speaking, and (2) Non-English speaking. The profile further indicated that the non-English speaking group was dominated by the Vietnamese and Cambodian community. Therefore, interpreters were employed to interview the Vietnamese and Cambodian irrigators. The procedure for household interviews is explained later in this chapter.

The sample size was based on the researcher's intention to include at least 50% of the irrigators who were using reclaimed water from the Virginia pipeline scheme. According to the WRSV sources (Collins 2005), the total number of irrigators using reclaimed water from the scheme was around 250 and 50 percent of these users comes to 125. However, given the limited resources (time, money and availability of the interpreters) along with factors such as irrigators' refusals to participate in the study, and unavailability of the irrigators for interview, the researcher found it more practical to visit as many farms as possible instead of taking the exact calculated figures. Accordingly, 165 farms were visited and the total sample size was 128 irrigators (includes both English speaking and non-English speaking communities). The total sample size for the household interview survey, and sample allocation are summarized in Table 5.1.

Table 5.1 Total sample size for household interview survey and sample allocation

Group	Total number of farms visited	Total respondents
Non-english speaking	120	91
Vietnamese	85	68
Cambodian	35	23
English speaking (Greek, Italian, Serbian, Turkish)	45	37
Total	165 (65.48)	128 (50.79)

Note Figures in parentheses are percentages of the irrigators (252) using reclaimed water from the scheme
Source Field survey

Willunga Basin Pipeline Scheme

Unlike the Virginia pipeline scheme, the Willunga basin pipeline scheme is entirely owned and operated by the growers, who have formed a private joint venture company. Most of the irrigators associated with this scheme are grape growers and/or wineries. For privacy reasons the Willunga Basin Water Company was not ready to supply the list of irrigators associated with the pipeline, which made obtaining the list of irrigators a difficult task. However, more attempts were made to convince the management of the Water Company about the purpose of the research and to seek their help in sending the letter of information. This time there was some success, as the management agreed to send the letters to its beneficiaries, but were doubtful about the response rate. The expectation was to receive consent from at least 20 irrigators (25% of total irrigators) to participate in the research. At the end of all this, 23 irrigators volunteered to participate in the telephone interviews. However, only 19 growers (24% of total growers) were interviewed, as the remaining four were not available for interview, despite repeated calls and messages left on their telephones. Table 5.2 provides the break up of the sample.

Musi River Basin

In case of the Indian case study, Water Users' Associations (WUAs) have been formed to manage the canals/tanks both downstream—where water flow is adequate due to wastewater inflows, and upstream—where there are serious water shortages. This has resulted in the upstream WUAs being silent passive bodies, with no active role in water management, while the downstream WUAs are more active. To make comparisons between them, the users' groups at both locations along the

Table 5.2 Total sample size for telephone interview survey at Willunga Basin pipeline

Particular	Number of growers
Total letters posted	80
Consent forms received	23 (28.75)
Total growers interviewed	19 (23.75)

Note Figures in parentheses are percentages of total irrigators (80)
Source Telephone survey

Table 5.3 Distribution of sample across the Musi River Basin

Particulars	WUAs downstream (with drainage)	WUAs upstream (without drainage)
Total number of WUA leaders interviewed	15	15
No. of Presidents interviewed	7	11
No. of TC members interviewed	8	4
Number of villages covered	15	15
Number of Mandals covered	5	3

Source Field survey

Musi River are included in the study. Accordingly, 30 WUA leaders: Presidents, or TC (Territorial Constituency) or DC members, were selected after discussions with key officials in the Irrigation Department and officials of the Institute of Resource Development and Social Management (IRDAS), Hyderabad, which also helped in conducting the interviews. Out of 30, 15 were located upstream and the other 15 were located downstream. Considering the time constraints it was decided to interview only Presidents of the WUAs. In their absence, any member of the TC was interviewed. Table 5.3 presents the distribution of the sample across the Musi river basin.

5.1.1.3 Selection of Key Stakeholders

Key stakeholders were selected for interview, with a focus on better understanding the workings of the schemes. The interviewees in Australia included people who were involved in the planning, operation and management of the reuse schemes and also represented different parties associated with in the scheme. In India, key stakeholders included officials from the Irrigation Department, researchers at the International Water Management Institute (IWMI) and people at local non-government organizations. The stakeholders were selected on the basis of their knowledge and perspective; the interviews were not highly structured and they took shape according to the individual's experience and time constraints.

5.2 Data Collection Methods

Both primary and secondary data were collected and used in the study. The secondary data included information, mainly from formal sources, such as SA Water, water companies and literature from a range of sources. The primary data sources included a household interview survey and interviews with key informants.

5.2.1   Household Interview Survey

The household interview survey consisted of face-to-face and telephone interviews with the irrigators associated with the reuse schemes. Face-to-face interviews were conducted in two cases—the Virginia pipeline scheme in Australia and the Musi irrigation scheme in India—while telephone interviews were carried out with irrigators associated with the Willunga pipeline scheme.

5.2.1.1   Face-to-Face Interviews

As observed during the initial exploration survey, irrigators associated with the Virginia pipeline scheme were from different ethnicities. For the sake of this study, these groups were broadly classified as English speaking and non-English speaking communities. Greeks, Italians and Australians constituted the English speaking communities while Vietnamese and Cambodians constituted the non-English speaking communities. Apart from the researcher himself, to gather survey data, eight interpreters (5 Vietnamese and 3 Cambodians) were recruited on the basis of their command of Vietnamese or Khmer and the English language, as well as relevant knowledge of the survey methods and study theme. The interpreters were further trained how to handle their jobs and the actual interviews were carried out under the researcher's supervision.

In case of the Musi irrigation scheme in India, trained interpreters who were also trained staff of a local non-governmental organization—the IRDAS in Hyderabad—were recruited for gathering household interview survey data. These interpreters were recruited on the basis of their command of the local language (Telgu), as well as of their relevant knowledge of the study area, survey methods and the study theme.

5.2.1.2   Telephone Interviews

Unlike the Virginia pipeline scheme, irrigators at the Willunga pipeline scheme came only from English-speaking communities, and hence it was decided to rely on telephone interviews with them. Professional interviewers at the Marketing Science Centre, University of South Australia, were employed for the interviews.

5.2.2   Stakeholder Interviews

Apart from the initial discussions held with the key stakeholders a second round of interviews were conducted at each scheme. The key stakeholders interviews were qualitative, in-depth interviews of knowledgeable sources, selected for their first-hand knowledge about each topic of interest (USAID 1996). The interviews were loosely structured, and relied on a list of issues to be investigated and hence allowed a free flow of ideas and information.

5.3 Survey Instruments

A structured interview survey questionnaire and a semi-structured interview questionnaire with sufficient room for probing, organized in a logical order of presentation, were used as instruments for data collection. The structured questionnaire was used for the household interview survey, while the semi-structured interview questionnaire was used for key informant interviews.

The household interview survey questionnaire was designed after considerable literature survey, consultation with key informants and local researchers. It used a mix of question types: multiple choice, numeric open-ended, rating scales and agreement scales. In the case of the Virginia pipeline scheme, the questionnaires were translated from English into Vietnamese and Khmer by professional translators to facilitate the interview process and, in some cases, to allow the respondents to easily mark the document themselves.

For the Musi irrigation scheme, the questionnaires were modified partly to satisfy the requirement of a larger research project and translated from English into Telgu by professional translators. This facilitated the interview process and also allowed the respondents to easily mark the document themselves, in some cases.

A shortened version of the questionnaire (with some irrelevant questions deleted) was used for the telephone interviews. A semi-structured interview questionnaire, with sufficient room for probing with second-order questions, was used for key informant interviews.

5.4 Data Analysis

The data gathered were analyzed in terms of the study objectives, and the analysis was carried out, using qualitative descriptions and descriptive statistics. The portion of data that is readily quantifiable (information from the close-ended questions of the questionnaire) was entered into the SPSS program and the output has been analysed using tabulations and cross-tabulations of variables, and with percentage values for the descriptive statistics. Readily non-quantifiable data (information from open-ended questions, key informant interviews, and focus group discussions) have been processed through qualitative description.

5.5 Methodological and Analytical Limitations

Although a proper attempt was made to collect the required field data there were some limitations. The sample size plays a significant role in interpretation and generalisation of the results. In the present study, the sample size in each of the case studies varied. For the Virginia pipeline scheme, a reasonable number of farm households (124) were interviewed. This accounted for 51% of the total number of irrigators associated with the VPS. For the Willunga basin pipeline scheme, the sample size was rather small (19) which accounted for 26% of the total number of irrigators associated with the scheme. Specific limitations for obtaining data have already been explained earlier.

The study depended mostly on the perception based qualitative information and therefore some limitations remain, such as (i) the nature of the information, (ii) its interpretation, and (iii) its suitability for statistical analysis. Further, the elicitation of subjective information based on the perceptions of the individuals is limited to the knowledge that they possess at the time of interview. However, it is hoped that the subjective information, which is based on a respondents' experience in the real world situation and their expectations of desirable change, would provide some insights, in the context of water governance in general and wastewater management in particular.

Since this study aims to examine the processes of governance and institution formation for urban wastewater management in Australia and India and therefore no data on the costs and benefits and economics of agricultural production were collected. So, willingness-to-pay and profitability analyses are beyond the scope of this study.

References

Buechler SJ, Devi G (2003) Household food security and wastewater-dependant livelihood activities along the Musi River in Andhra Pradesh. Report submitted to the World Health Organisation (WHO), Geneva, Switzerland, India

Collins John (2005) Personal communication. water reticulation services Virginia. Northern Adelaide Plains, Adelaide

Patton M (1990) Qualitative evaluation and research methods. Sage Publications, Newbury Park, California

Polit DF, Hungler BP (1999) Nursing research: principles and methods, 6th edn. J. B. Philadelphia, Lippincott Company, Philadelphia

Tellis W (1997) Introduction to case study. The qualitative report 3(2), July 1997

Trochim WM (2001) The research methods knowledge base (2nd edn). Retrieved from http://www.socialresearchmethods.net/kb/

United States Agency for International Development's (USAID) (1996) Center for Development Information and Evaluation. Conducting Key Informant Interviews. (Performance Monitoring and Evaluation TIPS). Retrieved from http://www.usaid.gov/pubs/usaid_eval/pdf_docs/pnabs541.pdf

van der Hoek W, Hassan MU, Ensink JHJ, Feenstra S, Raschid-Sally L, Munir S, Aslam R, Ali N, Hussain R, Matsuno Y (2002) Urban wastewater: a valuable resource for agriculture. A case study from Haroonabad, Pakistan. (Research report-63). International Water Management Institute, Colombo, Sri Lanka

Yin RK (1994) Case study research: design and methods 2nd edn. Applied social research methods series, vol 5. Sage Publications, California

Chapter 6
Public–Private Partnership Model for Wastewater Management

The first case studied is the Virginia pipeline scheme operating in the Northern Adelaide plains, South Australia. The scheme is the result of effectively designed partnerships and collective community efforts. In addition, effective regulatory and policy measures related to wastewater management in Australia, particularly South Australia have also been instrumental.

6.1 Background of the Virginia Pipeline Scheme (VPS)

The Virginia Pipeline Scheme is named after the township of Virginia, South Australia, which is the focal point of the Northern Adelaide plains. The region is described as South Australia's 'Vegie Bowl' because of its reputation for delivering high quality horticultural produce to local, interstate and overseas markets.

The scheme is built on the build-own-operate-transfer (BOOT) model, and is the largest of its type in the whole of Australia. It is a co-operative undertaking of the VIA, representing market gardeners and other irrigators; SA Water and WRSV (Water Reticulation Services Virginia), a private company. The proposal for developing the VPS was envisioned when the SA Water Corporation, as part of its Environment Improvement Program (EIP), constructed a filtration/disinfection plant (DAFF) costing AUD 30 million to treat lagoon effluent from the Bolivar wastewater treatment plant. This resulted in the production of Class A reclaimed water, which instead of being disposed of to the receiving waters, could be used for irrigation of the market gardens in the region, whose groundwater resources were already over-used. A private water company, WRSV, won a contract from the SA Water Corporation to access the output from the treatment plant, and also signed up clients for the reclaimed water and built the water distribution system. Since the project is built on the BOOT model, the project will be returned to the ownership of SA water by WRSV in 2019, at the end of the contract term. The total cost of the project (AUD 55 million), including the DAFF plant and the reticulation system,

© The Author(s) 2017
G. Keremane, *Governance of Urban Wastewater Reuse for Agriculture*,
SpringerBriefs in Water Science and Technology,
DOI 10.1007/978-3-319-55056-5_6

was shared between a Commonwealth Government contribution from the Building Better Cities funds (AUD 10.8 million); a Landcare contribution (AUD 574,000); private investors' contributions (AUD 7 million); SA government funds (AUD 7 million); the remainder was contributed by SA Water. As a result of the effective partnerships between the public and private entities, along with the collective efforts of the community, the scheme was finally commissioned in 1999, and since then has been operating successfully.

6.2   Results and Discussion

The data collected were analysed using SPSS software, and the results are presented in terms of frequencies and percentages.

6.2.1   Socio-Demographic Profile of the Respondents

Out of the total respondents interviewed, 89% were male while 10% were female. A majority of the respondents belonged to either the middle age group or the old age group. One interesting fact about this scheme is the diverse cultural backgrounds of the irrigators. However, as explained earlier, in this study they have been grouped simply as English speaking and Non-English speaking communities. The majority (71%) were in the non-English speaking community, which is also true for the total population in the region. The respondents were literate, as they all had attained education at no less than primary school level. Most of the respondents (47%) had farming experience, ranging from 6 to 10 years while around 5% of the irrigators had the experience of farming for more than 15 years.

6.2.2   General Awareness of Wastewater Usage and the Scheme

The respondents were asked whether they had any knowledge about reclaimed water use prior to the implementation of the scheme in order to examine the extent of knowledge possessed by the respondents regarding wastewater use. About 57% said they had some knowledge while 4% said they knew quite a lot. About 37% said they did not know anything about reclaimed water use. As most of the irrigators knew about wastewater use, clearly the community was aware that wastewater can be a useful and reliable alternative source of water to augment groundwater supplies, which is very important when implementing a water reuse scheme (MacDonald and Dyack 2004). In order to support this argument and check

Table 6.1 When did you first hear about the reclaimed water irrigation scheme?

Particulars	Percentage
Well before the implementation of the scheme (planning stage)	23.4
Just before the implementation of the scheme (implementation stage)	22.7
After implementation of the scheme (operational stage)	53.9

Source Field survey

whether the community was involved in the implementation of the scheme, the irrigators were asked about the time when they first heard about the scheme (Table 6.1).

Most of the respondents (54%) reported that they first came to know about the scheme after implementation, i.e., they joined the scheme once it had started operation. About 23% said they knew before implementation indicating that a considerable proportion of the respondents were involved right from the planning stage, which is once again a very important factor for any irrigation scheme to be successful, particularly wastewater reuse schemes.

The idea of community participation is being promoted and practiced in most parts of the world, particularly with respect to common pool resources management; the case with water reuse schemes in Australia is no different. Moreover, among other things, the National Water Initiative (NWI), signed by all Australian State Governments in 2004, promises better and more efficient management of water in urban environments through the increased use of recycled water and storm water. In addition, under one of the objectives of the NWI ('community partnership and adjustment'), the 'government are to engage water users and other stakeholders in achieving the objectives of the Initiative by improving certainty and building confidence in the reform processes; transparency in decision making; and ensuring sound information is available to all sectors at key decision points' (National Water Commission 2004, p. 20).

An important aspect of the VPS is the innovation with respect to the partnerships developed for achieving a common goal. It represents a case of well-designed 'public-private partnership' that has led to the success and sustainability of VPS. Following is a brief note on the framework of partnerships.

6.2.3 Framework of Partnerships

In the context of the water sector, a public-private partnership amounts to 'a public entity entering into a contractual agreement with the private sector to take over some or all of its activities related to water management' (OECD 2003). In general, public-private partnerships (PPPs) promoted within the water sector are concession-based contracts in which a private firm obtains from the government the right to provide a particular service under conditions of significant market power (Kerf et al. 1995, cited in Braadbaart 2005). Such contracts come in three flavours:

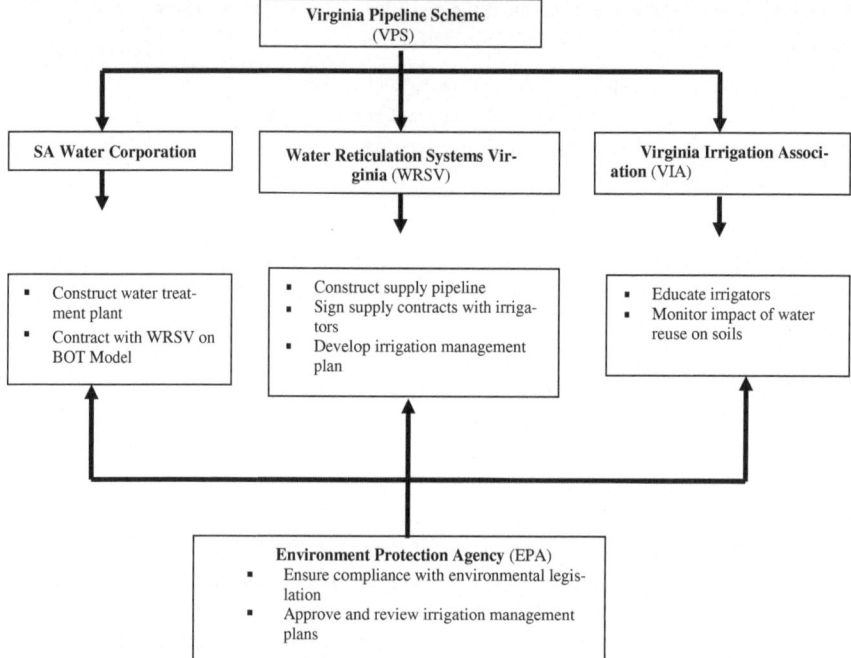

Fig. 6.1 Contractual agreements between key stakeholders in VPS. *Source* WRSV documents and Field survey

franchise contracts, concession contracts and build-own-operate-transfer (BOOT) contracts (Braadbaart 2005).

Implementation of the VPS was largely possible because of the enhanced participation of the stakeholders in effectively designed partnerships through contractual agreements between the stakeholders. As a part of the contractual agreement this scheme follows the Build Own Operate Transfer (BOOT) model. Figure 6.1 shows the contractual agreements signed by the stakeholders involved in the scheme.

In a BOOT project, a private company is given a concession to build and operate a facility, that would normally be built and operated by the government, and at the end of the contract period it is transferred back to the government (UNIDO 1996, cited in Braadbaart 2005). So in this case a private consortium (WRSV) is responsible for building and operating the Virginia pipeline scheme, until the whole scheme is returned to the ownership of SA water at the end of the BOOT period (Keremane and McKay 2007). Under this form of partnership, capital investment, designing and building, and operation of the scheme is the responsibility of the private sector, while the responsibility for setting performance standards, asset ownership, user fee collection, and oversight of performance and fees rests with the public agency; in the present case, SA water. The private company (WRSV) is responsible for designing, building and operating the scheme, as well as capital investment with contributions from SA water, State and the Federal governments in the proportions described at the beginning of this chapter.

To ensure that the irrigation of the agricultural land is sustainable, an Irrigation Management Plan (IMP) is developed. The responsibility for reporting deviations, if any, from the plan is assigned to WRSV. Ensuring that all environmental legislation is complied with is the responsibility of the Environment Protection Agency (EPA), which is also responsible for approving and reviewing the irrigation management plans on an annual basis. The irrigation association (VIA), representing the community/irrigators, is assigned the responsibility for managing an education programme for growers in relation to water reuse. Through this programme the VIA educates the irrigators about the impact of the enhanced nutrient levels on soils and natural groundwater due to the use of reclaimed water. It also closely monitors the effects of the reclaimed water on the soils. In addition, these arrangements also helped tackle the impediments—legal, policy, institutional, financial and social – that usually face the implementation of any reuse scheme.

6.2.4 Irrigators' Perception of Collective Action and Participation

The concept of collective action has emerged as a response to deal with the tragedy of the commons. We recall that the phrase 'collective action' refers to activities that require the coordination of efforts by two or more individuals (Agarwal and Ostrom 1999). Individuals associate in collective action to face uncertainties and search for solutions wherever possible. The commons literature has ample evidences of collective regulation for natural resources management (White and Runge 1995; Lam 1996; Ostrom 1992, 2000a). In the present case, an organised collective effort of the irrigators led to the implementation of VPS ultimately helping the irrigators to solve the problem of depleting groundwater resources.

Respondents were presented with scale items on collective action and participation and were asked to agree or disagree with these items. The results are presented in Tables 6.2 and 6.3. More than 75% of the respondents agreed that 'most people in the community are willing to help when in need'. When asked about their perception of community prosperity over the last five years, around 76% believed that the community had prospered because of cooperation among its members. Keeping in mind their variations in cultural background and ethnicity, the respondents were asked if they felt accepted as a member of a community. More than 70% agreed that they felt accepted. When specifically asked about cooperating during a water crisis, about 59% agreed that people cooperate in such situations.

The responses in general indicate that the community has a strong sense of cooperation and is community orientated. However, it is to be noted that a considerable percentage of people remained neutral (point five on the scale) in response to these propositions. Field observations matched the responses.

Table 6.2 Irrigators' perception about collective action and cooperation

Statements	Agree	Neutral	Disagree
People in the community will cooperate when there is water supply problem	58.6	14.1	27.3
Most people in the community are willing to help when in need	78.9	18.0	3.1
This community has prospered in the last five years	75.8	21.9	2.3
I feel accepted as a member of this community	77.3	20.3	2.3

Source Field survey

Table 6.3 Irrigators' perception about statements regarding participation

Statements	Agree	Neutral	Disagree
I have worked with others in the past for the benefit of the community	59.4	31.3	9.4
Most likely, the people who do not participate in communal activities are criticised	10.9	26.6	62.5
Everyone in the community make a fair contribution to communal activities	42.2	41.4	16.4

Source Field survey

Mere presence of a crisis does not always bring out collective action, participation of all the actors involved is equally important. According to Ostrom (2000b, p. 138), 'individuals in all walks of life and all parts of the world voluntarily organize themselves so as to gain the benefits of trade, to provide mutual protection against risk, and to create and enforce rules that protect natural resources'. This author, while highlighting the 'free rider' problem associated with collective goods, suggests that self-organized resource governance regimes can reduce its probability (Ostrom 1990, 2000b). Social cohesion, evidenced by a sense of community pride and identification, may convince individuals that working for a communal benefit is to their advantage (Meroka 2006) and therefore participation that addresses other factors that affect the likelihood of success is very important.

Participation is a broad term with many variations of meaning and interpretation. However, in its narrowest sense, it can be defined in terms of nominal membership, while in the broadest sense it can be defined as a process in which people have voice and influence decision-making (White 1996). Here, the focus is 'community participation' and to examine the extent of community participation the study proposed three statements to the irrigators (Table 6.3).

When asked if they worked with others for the benefit of the community, almost 60% of the respondents agreed with the proposition. When asked whether people who do not participate in communal activities are criticised, more than 60% disagreed. About making fair contribution towards communal activities around 42% of the respondents thought that everyone in the community did so (the term 'contribution' here meant contributing in terms of money or kind); almost an equal percentage of respondents remained neutral. Around 75% of the respondents agreed that the community had prospered over the previous five years.

Chi-square (χ^2) estimates were calculated to test whether irrigators' perception about collective action and participation varied with age, education level, or ethnicity. The estimates were not significant, confirming that irrigators' perceptions were similar across different age groups, education levels, and ethnicities.

Generally, reuse schemes span different agencies and, in this case too, the VPS is a co-operative undertaking and involves different agencies. Therefore, according to social capital theory, the irrigators' level of trust in these agencies is a very important measure of social capital which is an important ingredient of social sustainability.

6.2.5 Irrigators' Perception of Trust and Solidarity

Various agencies are involved in the functioning of the VPS, and trust in these agencies plays an important role in decisions about participation in the scheme. Respondents were asked about their level of trust in the agencies: government, EPA, health department and the water company. Figure 6.2 shows the level of trust that irrigators have in various agencies.

The irrigators had either complete trust or some level of trust in these agencies to perform their duties effectively. Around 58% of the respondents had complete trust in the government agencies, while another 16% had some level of trust. As for the water company, more than 55% had complete trust while around 26% more had some level of trust. About other associated agencies, like the EPA and the Health

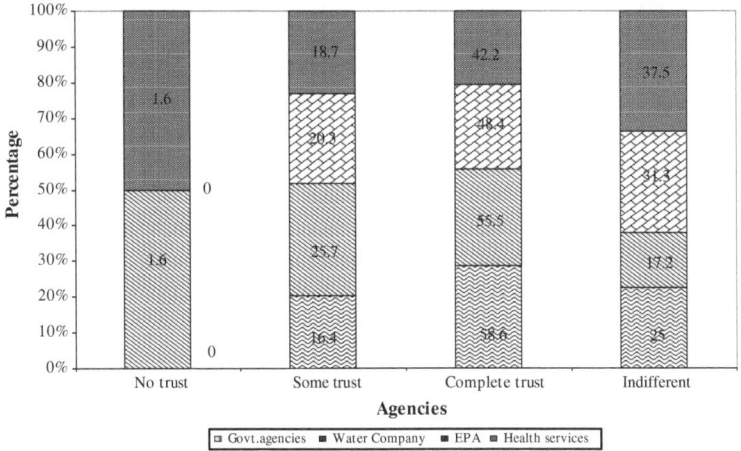

Fig. 6.2 Irrigator's level of trust in different agencies associated with the scheme. *Source* Field survey

Department, more than 40% had complete trust in them. However, the percentage of respondents who were indifferent is considerable, particularly with respect to EPA and Department of Health. This may be due the lack of awareness among the irrigators of the roles of these agencies in relation to the scheme.

The chi-square estimates for age group ($\chi^2 = 18.11$) and ethnicity ($\chi^2 = 41.78$) in respect of trust in the water company were significant, indicating that irrigators with English speaking background and in the young and middle age groups had more trust. Similar results were obtained in case of trust in the health department where the estimates for age ($\chi^2 = 23.82$) and ethnic groups ($\chi^2 = 71.32$) were significant.

It is evident from the success of the scheme that, despite different ethnicities and cultural backgrounds, the irrigators have demonstrated a high degree of networking; without this there might have been problems. This contradicts the argument on collective action that divisions between irrigators due to cultural and/or other social differences affect their capacity to communicate with one another (Tang 1992). Thus the findings of this study suggest that relatively heterogeneous community groups can be effective at provision of irrigation services (Kurian and Dietz 2005). It also demonstrates a high level of trust among the members of the community.

6.2.6 Irrigators' Perception of the 'Rules-in-Use'

Institutional arrangements are described using different terminologies by researchers studying common pool resources management and collective action (Tang 1992). However, in this case, we consider it to be the rules-in-use that stipulate who can participate in the scheme as appropriators and providers; what participants may, must or must not do; and how they will be rewarded or punished. These rules are conceptualised in the commons literature as "operational rules" (Tang 1992, p. 81). In order to elicit the perceptions of the irrigators about these rules-in-use they were presented with propositions and asked to indicate their degree of agreement with each of them. The responses are presented in Fig. 6.3.

When asked whether the rules governing water distribution were clear, around 60% of the irrigators agreed that the rules were clearly defined, with 34% strongly agreeing with this. About the process of water sharing or distribution within the scheme, more than 65% agreed that the process was appropriate and the results were similar when asked about the basis for allocating the water from the scheme, when more than 65% agreed that the allocation was fair. However, a significant number (31%) of the irrigators were neutral on this proposition. When they were asked about the water use charges and the basis of fixing them, most growers (50%) generally understood the price structure and were happy with the current price of the water (Marks and Boon 2005). Nevertheless, a significant percentage of the irrigators remained neutral. Perhaps this reflected their dissatisfaction with the 'take or pay' policy, as they were concerned about paying for an allocation whether or not they used the water.

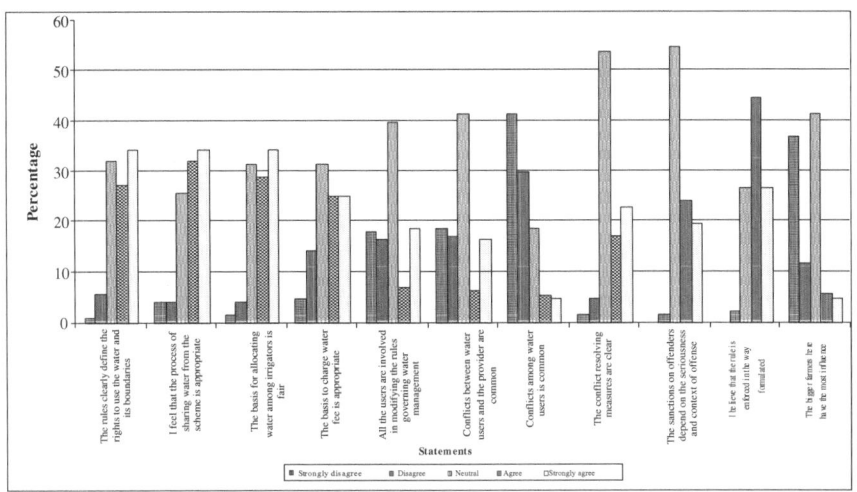

Fig. 6.3 Irrigators' perceptions about the operational rules in the VPS. *Source* Field survey

The survey went on to ask further whether all the irrigators were involved in decision making processes, particularly in modifying the rules governing the use of wastewater from the scheme. Around 34% disagreed with this proposition, stating that not all the irrigators were involved. It was observed that unlike some other self-governed institutions managing common pool resources, where the users create and modify the rules (Keremane and McKay 2006; McKay and Keremane 2006; Keremane et al. 2006), in this case the contractual agreement between the irrigators and the water company took care of these issues. This might have been the reason for the irrigators being neutral about water allocation and fees.

Generally, in natural resource management, conflicts arise due to disagreement over access, control and use of natural resources (Matriu 2000). It is more so with water because it has become a scarce resource in limited supply. So in this study the irrigators were presented with some propositions related to conflict and its management (see Fig. 6.3).

When asked if conflicts between the water company and the irrigators was common (common implying frequently occurring), around 41% of the irrigators remained neutral, around 35% disagreed, while about 22% agreed with the statement. This indicated some conflict; when the water company was asked about this, they said otherwise and also claimed there was no chance for conflict, as "everything is clearly mentioned in the contractual agreement and they adhere to it". Earlier results had shown that there was a strong sense of cooperation within the community; however the irrigators were asked if there were any conflicts with neighbours on water use; the results supported the previous observations, as 70% of the irrigators disagreed that there were conflicts between the water users. Although the water company had insisted there was no scope for conflict, the survey went on to ask the irrigators whether there were any conflict resolution measures mentioned in the agreement in case they should occur. More than 50% of them were neutral, indicating that they were not aware of any such measures. There was a similar response when they were asked about sanctions on offenders.

On a more general note, when the irrigators were asked if they believed that the rules were enforced as formulated, around 70% agreed that they were. Furthermore, over 40% felt that there was no influence from large farmers, which could be true, given that most of the farmers associated with the scheme were market gardeners.

6.3   Conclusions

Development of successful and sustainable water reuse projects will definitely provide solutions to water scarcity problems. However, we cannot overlook the impediments facing implementation of any reuse scheme. Conflicting agendas among water agencies; addressing water rights issues; dealing with opponents to recycling/reuse; modifying existing regulations and acquiring funding are some of the challenges to successful development encountered by reuse schemes.

Experience from the VPS suggests that, through collective action, enhanced community participation and well-designed partnerships, it is possible to coordinate individuals' activities; develop rules for resource use; impose sanctions on violators and mobilize the necessary financial, labour and material resources (Agarwal and Ostrom 1999). By providing knowledge and information on current best practice and communicating this information in a form that is understandable to the different stakeholder groups, it is possible to implement sustainable reuse schemes. This also influences the user's willingness to pay; the study found that willingness to pay for reclaimed water is influenced by various factors, such as the perceived benefits of the new facility, trust in the regulatory authorities, perception of ownership and understanding about the use and management of reclaimed water.

Fresh water scarcity and its associated problems are acknowledged world-wide. On the other hand, use of reclaimed or low quality water for potable and non-potable use has emerged as an innovative alternative option to augment continuously depleting freshwater supplies. However, for the latter option, use of this valuable resource imposes concerns about its suitability to sustain development, because of various issues related to wastewater usage and application. But as evidenced in the case of the Virginia Pipeline scheme, it can be said that by providing knowledge and information on current best practice, and communicating this information in a form that is understandable to the key stakeholder groups, any form of reuse can achieve sustainability, with its economic, social and environmental dimensions. Therefore, with sound policies, proper planning and management, sufficient financial commitments, and public awareness, support and participation it is possible to attain sustainability. Here are few suggestions from the VPS experience for the development of reclaimed water irrigation schemes in the future:

- Specific guidelines for wastewater use and management should be located and prepared.
- Awareness programmes regarding the legal, social, economic, environmental, and health issues related to waste water should target all key stakeholders.
- The private sector should play a key role in wastewater treatment and management.
- Enhanced community participation is crucial to achieving sustainability.

References

Agarwal A, Ostrom E (1999) Collective action, property rights, and devolution of forest and protected area management. Paper presented at Workshop on collective action, property rights and devolution of natural resources management. Puerto Azul, The Philippines, 21–25 June 1999

Braadbaart O (2005) Privatizing water and wastewater in developing countries: assessing the 1990s experiments. Water Policy 7:329–344

Keremane GB, McKay J (2007) Successful wastewater reuse scheme and sustainable development: a case study in Adelaide. Water Environ J 21(2):83–91

Keremane GB, McKay JM (2006) The role of community participation and partnerships: the Virginia pipeline scheme. Water 29(34):29–33

Keremane GB, McKay JM, Narayanamoorthy A (2006) The decline of innovative local self-governance institutions for water management: the case of Pani Panchayats. Int J Rural Manag 2(1):107–122

Kurian M and Dietz T (2005) How pro-poor are participatory watershed management projects?—An Indian case study. (Research Report 92). International Water Management Institute, Colombo, Sri Lanka

Lam WF (1996) Improving the performance of small-scale irrigation systems: the effects of technological investments and governance structure on irrigation performance in Nepal. World Dev 24(8):1301–1315

MacDonald DH, Dyack B (2004) Exploring the institutional impediments to conservation and water reuse—National issues. CSIRO Land and Water Client Report

Marks JS, Boon KF (2005) A social appraisal of the South Australian Virginia pipeline scheme: five years' experience. Report to Land & Water and Horticulture Australia Ltd., 20th May, Flinders University, Adelaide

Matriu V (2000) Conflict and natural resource management. FAO, Rome, p 1

McKay JM, Keremane GB (2006) Farmers' perception on self-created water management rules in a pioneer scheme: the Mula irrigation scheme, India. Irrigat Drain Syst 20:205–223

Meroka P (2006) New lessons for collective action: institutional change of Common Pool Resource (CPR) management in the Rufiji Floodplain, coast region Tanzania. Paper presented at the IASCP Europe Regional Meeting on Building the European Commons: from Open Fields to Open Source, Brescia, Italy, 23–25 Mar 2006

National Water Commission (2004) Intergovernmental agreement on a National Water Initiative–between the Commonwealth of Australia and the state governments of New South Wales, Victoria, Queensland, South Australia, the Australian Capital Territory and the Northern Territory. Retrieved from http://www.nwc.gov.au/nwi/docs/iga_national_water_initiative.pdf

Organization for Economic Co-operation and Development (2003) Public–private Partnerships in the urban water sector. OECD Policy Brief, April 2003

Ostrom E (1990) Governing the commons: the evolution of institutions for collective action. Cambridge University Press, Cambridge

Ostrom E (1992) Crafting institutions for self-governing irrigation systems. ICS Press, San
    Francisco
Ostrom E (2000a) Reformulating the commons. Swiss Polit Sci Rev 6(1):29–52
Ostrom E (2000b) Collective action and the evolution of social norms. J Econ Perspect 14(3):137–
    158
Tang SY (1992) Institutions and collective action: self-governance in irrigation. ICS Press, San
    Francisco
White S (1996) Depoliticizing development: the uses and abuses of participation. Dev Pract 6
    (1):6–15
White AT, Runge FC (1995) The emergence and evolution of collective action: Lessons from
    watershed management in Haiti. World Dev 23(10):1683–1698.

Chapter 7
Private Sector Participation in Wastewater Management

The Willunga pipeline scheme is the second reclaimed water scheme in Australia selected for the study. It is built by a joint venture company formed by grape growers and wine makers, which also owns and operates the scheme. Since inception, the scheme has been successfully supplying Class 'B' reclaimed water for growing grapes in the McLaren Vale region. The Willunga pipeline is a triple-bottom-line role model: it does not draw on public funds, it delivers high value to the community, and it reduces nutrient discharge to the ocean while replacing water consumption from aquifers and the river Murray. The scheme is an excellent example of how private sector participation backed up by favourable regulatory regime may well lead to solving water resource problems.

7.1 Project Background

The Willunga basin is home to the world-renowned McLaren Vale wine region and over 50 wineries. The McLaren Vale Wine Region is located just to the South of Adelaide, the capital of South Australia. However, during the mid to late 1990s the region missed out on the boom in wine exports because of dwindling water supplies, excessive groundwater extraction, and imposition of a water extraction licensing regime by the State government. The situation was that this prime grape growing region had ample land available, but no water to meet its irrigation needs (Gransbury 2004). Water had obviously become a scarce and valuable resource for the vineyards located in the basin, which had other associated problems such as declining crop yields and dropping land values. The situation demanded that the irrigators look for alternatives to augment the depleting fresh water supplies and this search led to the implementation of the Willunga pipeline scheme.

The Willunga pipeline scheme was commissioned in 1999, when the Willunga Basin Water Company (WBWC) negotiated a licensing agreement with the SA Water Corporation to access reclaimed water from the Christies Beach wastewater

© The Author(s) 2017
G. Keremane, *Governance of Urban Wastewater Reuse for Agriculture*,
SpringerBriefs in Water Science and Technology,
DOI 10.1007/978-3-319-55056-5_7

treatment plant for 40 years. The WBWC is a joint venture company formed by a consortium of grape growers and winemakers, which owns the pipeline and is responsible for its operation and maintenance. Unlike the Virginia pipeline scheme, this scheme did not receive any kind of financial support or subsidy from the public sector (State or Federal governments). All the costs incurred were met by the Water Company. Since the scheme started its operations it has benefited the company, community and environment.

One of the important drivers in initiating the scheme was the South Australian Environment Improvement Program (EIP) that completed in 2004. The EIP had the following aims: (1) increase the effectiveness of our metropolitan wastewater treatment plants (2) reduce the amount of treated wastewater entering Gulf St. Vincent and (3) recycle high quality treated wastewater for irrigation purposes. So when the Willunga Basin Water Company (WBWC) approached South Australian Water Corporation (SA Water) to gain permission to use the treated water from Christies Beach Waste Water Treatment Plant (CBWWTP) SA Water agreed to permit access for no charge. It was a win-win situation for SA Water, since without any investment it could comply with the EPA guidelines that wanted a reduction in the amount of treated effluent being discharged into the sea. Furthermore, the growers were able to get the alternative source of water that they were desperately seeking to expand their vineyards (WBWC 2006).

7.2 Private Sector Involvement and the Willunga Pipeline Scheme

According to Turrall (1995), ownership and management by profit-oriented companies, joint ventures, or non-profit organizations like user cooperatives, can all be included under the heading of privatisation. In this instance, the scheme under study is owned by a joint venture company—Willunga Basin Water Company (WBWC). According to Budds and McGranahan (2003, p. 90), 'a joint venture is an arrangement whereby a private company with the participation of private investors signs an agreement with the public sector whereby the private company takes a contract for utility management'. In case of the Willunga pipeline scheme, the WBWC (private company) formed by a consortium of grape growers and winemakers (private investors) signed an agreement with SA Water (public sector) to build, own, operate

Table 7.1 Allocation of responsibilities unde divestiture model in Willunga scheme

Responsibility	Sector
Asset ownership	Willunga Basin Water Company
Capital investment	Willunga Basin Water Company
Commercial risk	Willunga Basin Water Company
Operations/maintenance	Willunga Basin Water Company
Contract duration	40 years

Source Budds and McGranahan (2003)

and maintain the pipeline. The scheme has been a success since its inception, and this study attempts to examine the critical success factor for private sector participation, adopting a divestiture model. Table 7.1 presents the allocation of responsibilities under the divestiture model in case of the Willunga scheme.

7.3   Results and Discussion

The data collected was analysed using SPSS software, and the results were obtained in terms of frequencies, percentages and simple tabulations.

7.3.1   Socio-demographic Profile and Irrigation Details of the Respondents

Majority of the respondents (47.4%) belonged to 'old' age group and as for the respondents' education level, the survey indicated that the respondents were highly educated, since around 67% of the respondents had attained University degrees. For experience in farming, all the respondents had at least 10 years of experience in farming, and around 36% of the total respondents had more than 25 years of experience.

An inquiry of respondents' source of irrigation revealed that around 32% used a mix of groundwater, mains water/dam water and reclaimed water for irrigation purposes, while around 32% used only reclaimed water for irrigation. Twenty one percent of the respondents used a combination of reclaimed water and groundwater for irrigation purposes, while around 16% used a mix of mains/dam water and reclaimed water. Since a significant amount of reclaimed water was being used for irrigation, the survey went on to ask the respondents about the proportion of reclaimed water in the total water used to irrigate vineyards.

Around 37% said that reclaimed water bought from the company was more than 75% of the total water used. Out of this, around 31% used 100% reclaimed water to irrigate their vineyards. About 32% said they used less than 50% of reclaimed water while a similar percentage of respondents said that the proportion of reclaimed water being used ranged between 50 and 75%. In terms of acreage (area irrigated using reclaimed water), around 42% of the respondents used reclaimed water to irrigate up to 20 acres of their land, while about 37% had more than 45 acres of land under reclaimed water irrigation.

It is clear from the responses received that reclaimed water is a major source of irrigation for the respondents in the study area. According to a recent report, the use of ground water in the McLaren Vale area has declined by more than two-thirds in just over a decade. A trend of wetter spring months, along with changing land use and irrigation practices, maturing vines and greater use of reclaimed water is credited for the decline.

7.3.2   Reason for Implementation
of the Scheme and General Awareness

Many factors as discussed earlier have led to the development of the Willunga pipeline. However, from irrigators' point of view groundwater depletion was the major driving force behind the implementation as indicated by majority (73.6%) of the growers. There were other reasons too that contributed to the start of this scheme, such as the high price of mains water, encouragement by water authorities, and community interest. An important factor in the success of any water reuse scheme is community involvement and awareness. This is evident for the Willunga pipeline scheme, in the sense that more than 50% of the respondents had been involved with the scheme since the planning stage (Fig. 7.1). Around 36% of the respondents were involved during the implementation stage, while 10% joined during the operational stage. In addition, it was noticed that the respondents had some knowledge (68.4%) or knew quite a lot (31.6%) about water reuse even before the scheme was implemented.

A query about the source of information further supported this observation about the knowledge and awareness among the community. It was observed that general knowledge (47.4%), followed by the personal experience of the growers (21%), were the major sources of information. These results indicate that the community was aware of the advantages of reclaimed water reuse in general, and of the scheme in particular. Further, it also shows that the community was involved right from the planning stage in the development of the Willunga pipeline scheme. Thus, it was these realisations by growers and the interest of SA Water Corporation in preventing discharge of nutrients to the bay that led to implementation of the scheme.

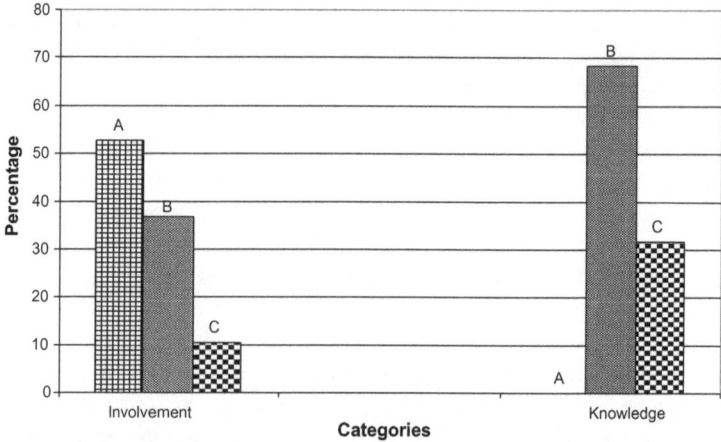

Fig. 7.1 Irrigators involvement and level of knowledge. *Notes A, B, C* for Involvement indicates three stages: planning, implementation, and operational. *A, B, C* for Knowledge indicates three levels: no knowledge, some knowledge, and quite a lot. *Source* Field survey

Over and above all these factors, a favourable regulatory and policy regime for wastewater in South Australia and several aligned regulations, such as reclaimed water use guidelines, the approval of Public and Environmental Health and the EPA were also instrumental in the commencement of the scheme (McKay 2007).

7.3.3 Irrigators' Knowledge of the Operational Details of the Scheme

The survey went on to investigate the irrigators' perceptions about the operational details of the scheme: ownership of the scheme, the authority for deciding the water charges and accessibility to the scheme, and the operation and maintenance of the scheme (Fig. 7.2). The irrigators were agreed in their responses. About ownership, around 89% thought that the scheme is owned by the Water Company, 5.3% of the irrigators believed the State government owned the scheme, while an equal percentage had no idea.

The results were quite similar when the irrigators were asked: *Who has the authority to decide about accessibility to reclaimed water from the scheme?* All the irrigators (100%) were of the opinion that the water company had the authority to decide upon the water charges and was entrusted with the responsibility for the operation and maintenance of the scheme.

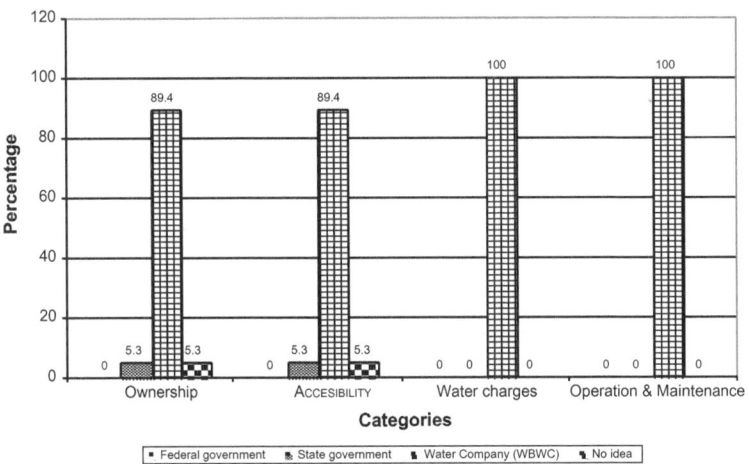

Fig. 7.2 Irrigators' perceptions about the operational details of the scheme. *Source* Field survey

7.3.4   Irrigators' Knowledge and Level of Trust

The survey tried to examine the extent of knowledge that the irrigators and the water company had regarding wastewater treatment and usage, since this is important when it comes to community involvement. To a certain extent it also implies the level of trust the users have in the resource (reclaimed water) and the water providers' ability to deliver the goods. Figure 7.3 presents the perceptions of the irrigators.

Irrigators were asked about the knowledge they had regarding wastewater treatment and usage. The results indicate that more than 90% had this knowledge. However, the extent of their knowledge varied, as a majority of the irrigators (47%) knew quite a lot about wastewater treatment and usage, while around 15% said they were fully informed. About 31% said they had some knowledge, while a small number (5%) of the irrigators said they had no knowledge at all.

When they were asked about the extent of knowledge that the Water Company had regarding wastewater treatment, all the irrigators (100%) said the water company was knowledgeable. More than 50% said it was fully informed, around 31% said the water company knew a lot about wastewater treatment and usage, and about 15% said the water company had some knowledge.

The results clearly show that the community is aware of wastewater treatment and usage. Also, the community has full faith in the water company to deliver the goods, as it perceives the Water Company to be knowledgeable about the processes, a very positive sign for the success of any reuse scheme. However, the survey tried to ask some direct questions to examine the level of trust the community (irrigators) had in the different agencies that are involved, directly or indirectly, with the

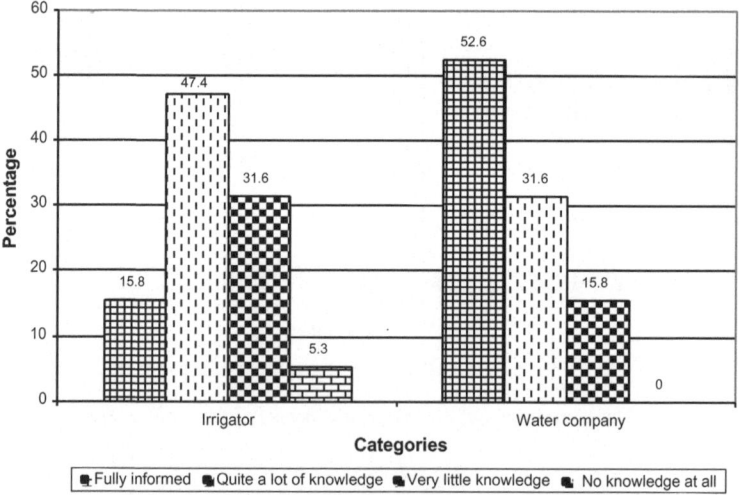

Fig. 7.3 Irrigators' and water company knowledge of wastewater treatment and use. *Source* Field survey

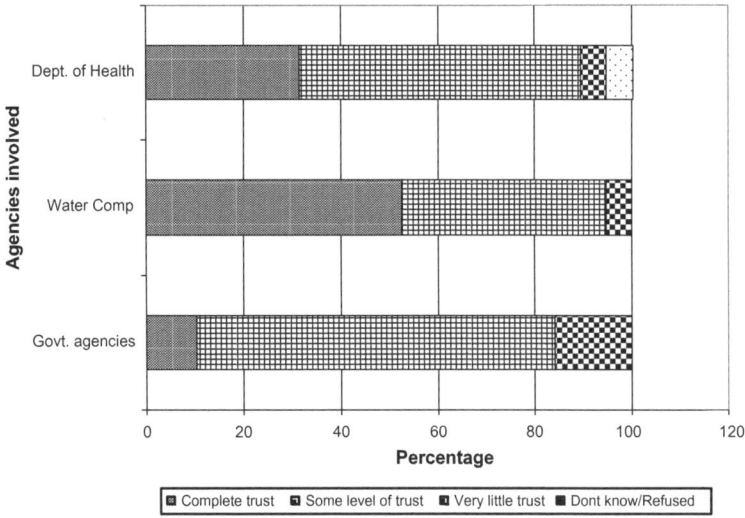

Fig. 7.4 Irrigators' level of trust in various agencies associated with the scheme. *Source* Field survey

scheme—the Government agencies, Water Company, and the Department of Health. Their responses are presented in Fig. 7.4.

Generally, the irrigators had trust in all the agencies involved with the scheme. However, the level of trust was higher in the Water Company, compared with the government agencies and the Department of Health. As illustrated in Fig. 7.4, around 70% of the irrigators said they had some level of trust in the government agencies, and only 10% said they had complete trust; about 15% said they had very little trust. In the case of the Water Company, more than 50% of the irrigators said they completely believed the Water Company would perform its duties, and around 42% said they had some level of trust in the water company. As any reuse schemes raises health concerns, trust in the agencies involved in defining the health requirements (Department of Health in this case) is equally important. When asked about this, more than 55% of the irrigators had some level of trust in the department, while 31% had complete trust; the least number of irrigators (5%) had no trust. An equal number of the irrigators refused to answer (Fig. 7.4)

Further, 'trust' is one of the most frequently encountered elements in definitions of social capital (Hutchinson and Vidal 2004) and maintaining social capital means social sustainability (Keremane and McKay 2006). According to Goodland (2002, p. 490), 'social capital is investments and services that create the basic framework for society. It lowers the cost of working together and facilitates cooperation'. In this case, it is important to note that the Water Company is a consortium of the grape growers and the wine makers who are the 'community' in the study region, and hence there is high level of trust within the community. Thus, the presence of high social capital within the community is one of the factors contributing to the successful functioning of the Willunga scheme.

7.4 Conclusion

Wastewater use in agriculture will definitely ease the pressure on available freshwater resources. But, successful development of reuse schemes encounters social, institutional, financial, regulatory, and technical impediments. Overcoming these hurdles will lead to implementation of a successful reuse scheme as evident from the present case study. 'How' is the question to be answered? An examination of the Willunga pipeline scheme revealed that six factors: *social, institutional, financial, regulatory and policy, risk allocation, and technical* are critical in implementing water reuse scheme trhiugh private sector participation. Also, the case of Willunga pipeline scheme illustrates that considerable innovation and collective effort by a motivated group of individuals can yield positive results. Finally, the findings provide insights to develop similar schemes in the developing world, where private sector participation is in its infant stage provided the following areas receive proper attention:

- Appropriate arrangements and agreements between the government, the management company and the growers;
- Thorough financial and technical feasibility studies to ensure the scheme's long-term viability and to attract private sector funding;
- Competent technical design and appropriate safety measures and practices to avoid any occupational health and safety hazards;
- Regular water quality monitoring and control, best irrigation practice through soil surveys, regular soil and crop management reports, in order to ensure environmental sustainability;
- A tariff structure that is affordable to the users and also ensures the financial sustainability of the project;
- Above all, a favourable policy and regulatory regime for wastewater reuse, based on the local socio-economic and political situations.

References

Budds J, McGranahan G (2003) Are the debates on water privatization missing the point? Experiences from Africa, Asia and Latin America. Environ Urbanization 15:87–113

Goodland R (2002) Sustainability: human, social, economic and environmental. In: Timmerman P (ed) Social and economic dimensions of global environmental change, encyclopedia of global environmental change, vol 4. Wiley

Gransbury J (2004) The willunga basin pipeline. Plenary presentation at the irrigation Australia 2004 conference, The Irrigation Association of Australia Ltd, New South Wales

Hutchinson J, Vidal AC (2004) Using social capital to help integrate planning theory, research, and practice. J Am Plan Assoc 70(2):142–192

Keremane GB, McKay JM (2006) The role of community participation and partnerships: the Virginia pipeline scheme. Water 29(34):29–33

McKay JM (2007) Policy changes in South Australia: elements of the social contract resulting in high urban and agricultural use of recycled water. South Australia Policy Online. Retrieved 13 June 2007 from http://www.sapo.org.au/opin/

Turrall HN (1995) Recent trends in irrigation management changing directions for the public sector. Nat Resour Perspect 5

Willunga Basin Water Company (2006) Personal communication: meeting with the Willunga Basin Water Company management-norm Doole, Kym Davey, and Glen Templeman, 22 March 2006

Chapter 8
Informal and Uncontrolled Use of Wastewater for Agriculture

The third case study is in India, where wastewater use for irrigation is unregulated and indirect, similar to the practices in many other developing countries. Wastewater reuse is not new to India, because there has been a history of untreated or partially treated wastewater use there for a long time. Today, as a result of rapid population growth, massive industrialization, and the growing number of cities, indirect use of wastewater has increased even further as large amounts of sewage are discharged into the rivers. Most of this reuse occurs along the Indian peninsular rivers for agricultural irrigation and it is important to note that these rivers would not have had any flow for most of the year if they were not used to funnel wastewater away from cities to peri-urban and rural areas (Buechler et al. 2002; Buechler 2004). The Musi River, which rises a few kilometres upstream from Hyderabad and flows across Telangana (erstwhile Andhra Pradesh) is one of these many rivers (Buechler and Devi 2003).

8.1 Field Settings

The river Musi spreads over 8000 km^2. and lies in a region receiving an annual rainfall ranging from 500 to 700 mm. The basin is drained by many small streams, and most of the water flows are diverted to a series of tanks and used for irrigation. However, these inflows are very limited; further on, as the river passes through the city of Hyderabad (Capital of Telangana State), with a population of around seven million, the wastewater from the drains passes into the Musi, almost throughout the whole year. Wastewater released into the river is untreated or partially treated; most of it is released from the two wastewater treatment plants operating in the region. One of these plants has primary and secondary treatment, while the other just has primary treatment facilities. According to estimates, only 40% of the sewage is clarified before it is dumped into the river (Buechler et al. 2002).

© The Author(s) 2017
G. Keremane, *Governance of Urban Wastewater Reuse for Agriculture*,
SpringerBriefs in Water Science and Technology,
DOI 10.1007/978-3-319-55056-5_8

The field situation is that, upstream of the place where the river enters the city, it has no water in it, except during the monsoons; while downstream, due to the discharge of vast amount of wastewater the river is perennial. The quantity of wastewater released into the Musi is estimated to be around 5200 litres per second (IRDAS 2006).

8.1.1   Channeling of Wastewater for Irrigation

As explained earlier, the irrigation schemes along the Musi River depend primarily on urban wastewater from the city of Hyderabad, and these schemes are controlled by the Irrigation Department. However, the storage and channelling of wastewater for use in agriculture varies, and is diagrammatically represented in Fig. 8.1.

Most of the wastewater discharged into the Musi (around two-thirds of the total discharge) is channelled via open sewage drainage canals. The remaining one-third is channelled through the sewage system to either of the two treatment plants, from where the partially treated wastewater is channelled downstream via a canal used for agricultural irrigation. In some cases the wastewater from the sewage treatment plant is stored in a natural pond where the untreated and treated water mix. The water is then pumped from the pond and used for irrigation.

The channelling methods used to irrigate the lands along the river vary depending on their location. Generally, wastewater from the river is first diverted via anicuts (local name for weirs) on both sides of the river to main canals that further feed the branch canals (Fig. 8.2). There is direct irrigation from the branch canals or main canals in the case of the fields closest to the riverbanks. For fields

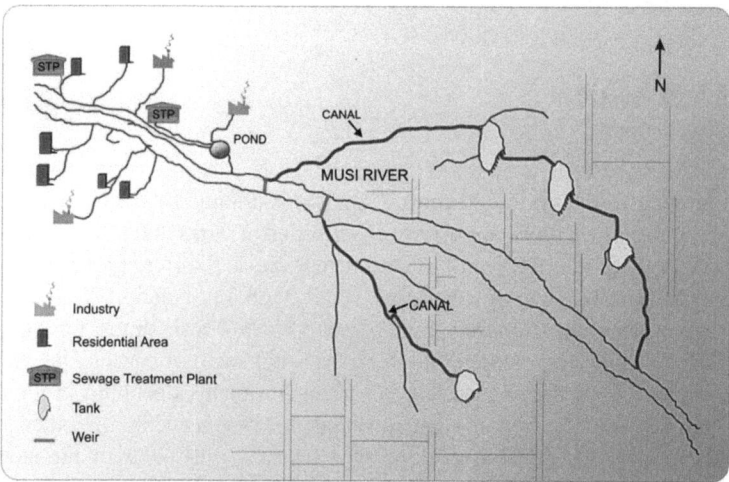

Fig. 8.1 Methods of channelling wastewater for irrigation along the Musi River. *Source* Modified from IWMI (2007)

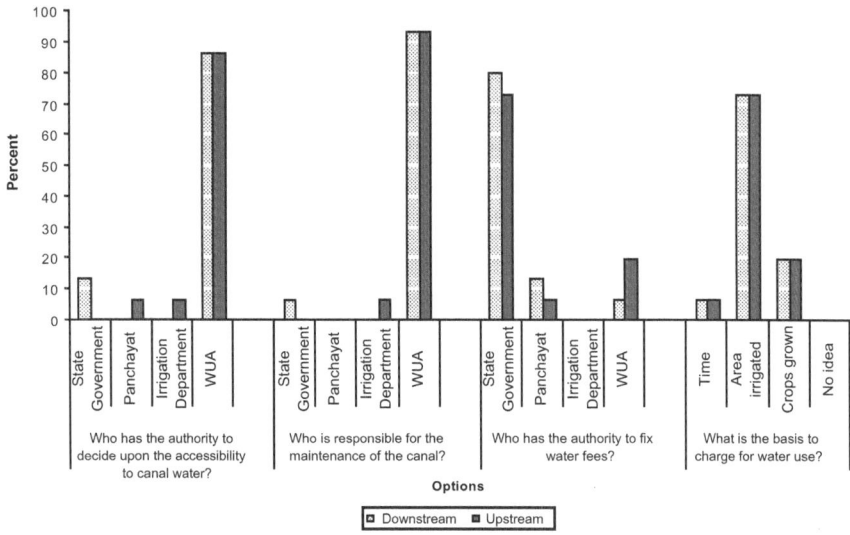

Fig. 8.2 WUA leaders' perceptions about water management issues. *Source* Field survey

located at higher altitudes, wastewater is pumped from the branch canals into underground pipes and later directed to smaller channels that go to the fields. In some other cases, the water from the weirs is channelled to tanks of varying sizes, where it is stored for irrigating the fields near the tanks. These tanks are controlled by the Water Users Associations (henceforth WUAs) formed under the Andhra Pradesh Farmers Management of Irrigation System Act (APFMIS) enacted in 1997 (see Box 8.1). This study focuses on these WUAs that were formed originally to manage surface irrigation systems, whereas at present they are managing the use of wastewater downstream of the river Musi.

Box 8.1 The Andhra Pradesh Farmers Management of Irrigation System Act, 1997

The State of Andhra Pradesh (now Telangana) by passing the Andhra Pradesh Farmers Management of Irrigation Systems (APFMIS) Act in 1997, laid pathways to irrigation sector reforms in India. The APFMIS Act facilitates: (a) formation of WUAs (Water Users' Associations) on the basis of a hydraulic boundary; (b) the inclusion of landowners and tenants; (c) making a person eligible to become a member of more than one WUA; (d) the exclusive right of members to vote (either owners or tenants). The Act has provisions for the election of president and members of the managing committee for a period of three years at three levels: (i) WUA level, (ii) distributory level, and (iii) project level. The APFMIS Act has clearly underlined the objectives, functions and resources of WUAs. The Act also identifies the specific responsibilities and tasks of government officials and WUA leaders. The Act transfers control over field personnel of the state Irrigation Department to WUAs and makes membership in primary-level WUAs compulsory, along with obligations of membership, including fee payment. The Act requires annual budgets of WUAs be brought before the general body of the WUA for approval. The Act gives WUAs legal personalities and powers, including the right to levy taxes and impose fines, which are ultimately enforceable through the legal powers of the state. The Act separates WUAs from the local political establishments and allows the government to resume either governance or operational control from WUAs in the event that they fail to perform effectively.

Sources Sivamohan (2001), IRDAS (2006)

However, it is important to note that unlike the previous two cases which represented planned and regulated wastewater reuse, the situation with the Indian case study is different, where wastewater use and management is largely unplanned and unregulated. Further, field observations and discussions with the Irrigation Department officials revealed that in India, different informal institutions and organizations are associated with wastewater use at different levels—macro, meso and micro. Accordingly, in Hyderabad, which is the study site, there were a number of informal institutions related to wastewater use along the river Musi. In urban areas, there was an Urban Farmers' Association, primarily composed of wastewater farmers who own land, while in the peri-urban and rural areas it was the WUAs that were composed of farmers with access to wastewater. This study focuses the community organisations or the Water User Associations which are operating at the micro-level.

8.2 Results and Discussions

The WUAs along the Musi with wastewater inflows are comparatively old, as a couple of WUAs were formed in 1995 and 1996, even before the passing of APFMIS Act in 1997 (Table 8.1). These WUAs were located in two districts, Nalgonda and Rangareddy, and they represented different types of irrigation system: tank, canal or nala as indicated by their names. The majority of the WUAs (7) downstream were located in the middle reach of the Main canal while four each were located on the head and tail reaches. On the distributory, six WUAs each were on the head and middle reaches respectively, and three were located on the tail reach. The WUAs formed upstream are comparatively new, as all these were formed following the passing of the APFMIS Act (Table 8.1). They were all concerned with canal irrigation and belonged to the Nalgonda district.

Table 8.1 Profile of the WUAs selected at upstream and downstream of Musi River, receiving wastewater inflows

Upstream WUAs				Downstream WUAs			
Name of the WUA	Year formed	Location of the WUA		Name of the WUA	Year formed	Location of the WUA	
		On main canal	On distributory			On main canal	On distributory
Pedda Cheruvu WUA	2001	Head reach	Head reach	Chowdari Kathwa WUA	1997	Middle reach	Head reach
Ura Cheru WUA	2001	Tail reach	Head reach	Laxminarayana WUA	1999	Tail reach	Middle reach
Pedda Cheruvu WUA	1999	Tail reach	Head reach	Bacharamkalwakatwa WUA	1999	Middle reach	Middle reach
Pedda Cheruvu WUA	2001	Head reach	Head reach	Gowrelly WUA	1999	Middle reach	Tail reach
Pedda Cheruvu WUA	1999	Tail reach	Middle reach	Yerrakunta WUA	1997	Middle reach	Head reach
Pedda Cheruvu WUA	1996	Head reach	Middle reach	Pillayapally Nala	1998	Middle reach	Middle reach
Pedda Cheruvu WUA	1997	Head reach	Middle reach	Sangem Anikut WUA	1998	Middle reach	Head reach
Thumma Cheruvu WUA	1997	Tail reach	Tail reach	Dharma Reddy Pally Kathwa	1996	Head reach	Head reach
Akkamatti Cheruvu WUA	1997	Head reach	Head reach	Ramasamudram Chervu WUA	1996	Middle reach	Middle reach
Veerla Cheruvu WUA	1997	Head reach	Tail reach	Asif Nahar channel WUA	1998	Tail reach	Middle reach
Mogulla Cheruvu WUA	1997	Middle reach	Middle reach	Akkachellalla Cheruvu WUA	1996	Tail reach	Head reach
Pedda Cheruvu WUA	2000	Head reach	Head reach	Bapanenu Cheruvu WUA	1999	Tail reach	Middle reach
Pedda Cheruvu WUA	1997	Tail reach	Middle reach	Rudravelli Cheruvu	1995	Head reach	Tail reach
Dacharam WUA	2000	Head reach	Head reach	Alinagar channel WUA	1998	Head reach	Head reach
Pedda Cheruvu WUA	1997	Head reach	Head reach	Chintala Cheruvu WUA	1996	Head reach	Tail reach

Note Cheruvu means tank in Telugu (local language). *Source* Field survey

8.2.1   Socio-demographic Profile and Irrigation Details of the Respondents

Since the study aimed at looking into the perception of the irrigators across the WUAs located upstream and downstream of the river Musi, the results are presented accordingly. The figures indicate that irrespective of the location, the majority of the respondents (around 60% in both cases) belonged to the older age group. Almost all the respondents were literate, with the majority (about 40% in both cases) of the respondents having primary education. Regarding farming experience, the percentage varied across the location. In the downstream section, where the river was perennial due to wastewater availability, a majority (60%) of the respondents had less than 25 years of farming experience, while a majority (80%) of the respondents upstream were engaged in farming for more than 25 years. This is because most of the farming downstream is along the peri-urban area and this picked up recently, with more sewage being discharged to the river Musi as a result of urbanisation and industrialization.

Although the main source of irrigation is canal water, the farmers in the region depend on other sources as well, as indicated in Table 8.2. General observation and discussions with farmers revealed that generally the farmers upstream did not receive enough water from the canal, especially during summer. However, downstream, the case was different due to the sewage water inflows.

Wirh regards to the source of irrigation, majority of the respondents downstream (53.3%) were using both canal water and wastewater, while upstream around 47% used groundwater and open wells to meet their irrigation needs.

8.2.2   Perceptions of WUA Leaders About Water Management

As discussed earlier, the WUAs under study were not formed to manage wastewater in particular. These were the WUAs formed under the APFMIS Act of 1997. It is

Table 8.2 Distribution of respondents based on source of irrigation

Particular	Number of respondents	
	Musi river downstream	Musi river upstream
Canal water only	26.7	0.0
Groundwater only	0.0	26.7
Other sources (open well)	0.0	13.3
Canal + groundwater	0.0	13.3
Canal water + wastewater	53.3	0.0
Canal + groundwater + wastewater	20.0	0.0
Groundwater + other source (open well)	0.0	46.7

Source Field survey

important to note that in the study region, particularly downstream of Musi where the river is perennial, we can find other informal associations or users groups, such as the Urban Farmers' Association, different caste (low caste) groups in addition to the WUAs formed under the Act. However, this study focuses on the WUAs that were formed as a result of irrigation reform strategy in Andhra Pradesh.

The study interviewed the WUA leaders, who were either the President of the WUA under study or the Territorial Constituency (TC) Members. The idea was to understand the perceptions of these leaders regarding issues related to water management in their respective WUA. Figure 8.2 presents the perceptions of WUA leaders about issues such as operation and maintenance, water fees, and accessibility to water from the canal. The responses to the questions related to water management were similar at both locations, downstream and upstream.

When asked who has the authority to decide upon the accessibility to canal water, more than 85% at both locations said it was the WUA's decision. Around 13% of the respondents representing the WUAs downstream of Musi said that the State government decided about accessibility. The respondents from WUAs located upstream believed it was the Irrigation Department (6.7%) and the Panchayat (6.7%) who decided about accessibility to canal water. More than 90% of the respondents at both locations believed that the WUA was responsible for the operation and maintenance of the canal.

The study went on to ask the respondents what they felt about the rules governing water use in their area. The leaders were presented with propositions representing various aspects of water distribution and rules within the WUA under study and they were asked to agree or disagree with the propositions.

The scale items presented to the WUA leaders represented the normal activities of a WUA, including statements mainly related to operation and maintenance, and to the formation and performance of the Executive Committee (Table 8.3).

When asked if the water sharing among all the irrigators was fairly done, all the downstream WUA representatives (100%) agreed that it was fair, while around 40% of the upstream representatives thought it was unfair. When asked whether the water distribution system was efficient, again the response was on similar lines. A majority downstream (80%) agreed that the system was efficient, while more than 50% of respondents upstream disagreed. While 80% of respondents downstream felt secure with the present water distribution system, more than 65% felt they were not secure. Furthermore, more than 50% of respondents upstream disagreed that the basis for distributing water was appropriate, while it was the reverse downstream, with more than 65% agreeing with this statement.

However, the responses to statements representing the general administrative aspects of WUAs were almost alike at both locations. A majority of the respondents at both locations agreed that the Executive Committee was formed fairly and that it was fair in its processes (Table 8.3). They also agreed that there was not much discrimination based on caste or land holding. This is mainly because the majority of the respondents (83%) belonged to the backward community and were marginal farmers (Buechler et al. 2002).

Table 8.3 WUA leaders' perceptions of rules for water distribution

Scale items	Upstream			Downstream		
	Disagree	Agree	Neutral	Disagree	Agree	Neutral
Water is shared fairly among every user in the scheme	40.0	40.0	20.0	0.0	100.0	0.0
Water distribution system is efficient	53.3	33.3	13.3	6.7	80.0	13.3
I feel secure with the present water distribution system	66.7	20.0	13.3	13.3	80.0	6.7
The basis for distributing water among irrigators is fair	53.3	20.0	26.7	26.7	66.7	6.7
The basis to charge water fee is appropriate	6.7	53.3	40.0	0.0	86.7	13.3
The way executive committee is formed is fair	6.7	80.0	13.3	13.3	66.7	20.0
The committee is fair in its processes	13.3	80.0	6.7	13.3	73.3	13.3
All caste members get an equal hearing during meetings	0.0	100.0	0.0	0.0	100.0	0.0
Bigger farmers are more influential	93.3	6.7	0.0	100.0	0.0	0.0
The rules are enforced as formulated	40.0	46.7	13.3	40.0	60.0	0.0

Source Field survey

The results clearly indicate that, compared to the respondents who represented the WUAs downstream, their counterparts upstream tend to disagree more with the proposed statements, meaning that the percentage of respondents disagreeing with the proposed statements was comparatively higher in the case of the WUAs located upstream. This is mainly because, upstream the Musi River runs dry throughout the year, except during monsoons, in contrast to downstream, which receives wastewater all year round. Hence the downstream WUAs performed better than those upstream, which is reflected in the responses of the WUA leaders. This also proves that wastewater is an important substitute source for meeting declining freshwater resources.

A very important issue in natural resource management is conflict over access to, control and use of natural resources and their management. In the case of irrigation systems, particularly in India, these conflicts arise over: unauthorized use of water, breaking the rotational sequence, illegal use of water, and wastage of water (McKay and Keremane 2006). The study examined this aspect of water management by self-governed institutions in the study region and Table 8.4 presents the results.

The respondents were asked whether conflicts between water users and the executive committee was a common phenomenon. The word 'common' was used in the study to denote the frequency of the conflicts occurring at a given period of time. It was interesting to note the differences in opinion at each location. More than 70% of the WUA representatives upstream disagreed that conflicts were common; while downstream 33% agreed that conflicts were common. When the respondents were asked about conflicts among water users, around 20% of the respondents upstream agreed that there were conflicts, while only 6% did so downstream. These results highlight two very different yet interesting aspects of water management.

The fundamental situation is that there is less water upstream than downstream. Therefore, judging by the responses of the WUA leaders at each location, it is evident that on the one hand, 'less water—fewer conflicts between the irrigators and the WUA' while on the other hand, 'there can be conflicts between users over limited water resources'. Nevertheless, it is clear that there were conflicts, as there are bound to be with natural resource management (Matriu 2000); so two more statements related to conflict resolution were presented to the interviewees (see Table 8.4).

When asked if the conflict resolution mechanisms were clear and in place to resolve any conflicts that might occur, around 40% of respondents representing upstream WUAs disagreed, while for downstream WUAs, more than 70% agreed with the statement. Subjecting the defendant to social pressure was found to be a common procedure employed to resolve conflicts among the WUAs in Maharashtra (Keremane and McKay 2006; McKay and Keremane 2006). Therefore, the respondents were asked if 'social pressure' was a common method of resolving conflicts. It appeared that the case was similar in the study region as well, since a majority of respondents in each location agreed with this statement.

Table 8.4 WUA leaders' perceptions about cooperation, conflicts and their resolution

Scale items	Upstream			Downstream		
	Disagree	Agree	Neutral	Disagree	Agree	Neutral
Conflicts between water users and the management are common	73.3	6.7	20.0	53.3	33.3	13.3
Conflict between members is common	40.0	20.0	40.0	80.0	6.7	13.3
In case of any conflicts, the conflict resolution measures are clear and in place	40.0	26.7	33.3	26.7	73.3	0.0
Social pressure is the common conflict resolution mechanism practiced	40.0	46.7	13.3	40.0	53.3	6.7

Source Field survey

8.3   Conclusions

In many developing countries such as India, the rapid expansion of urban development will bring opportunities in terms of increased water supply for irrigation in the form of wastewater. This is exactly the situation in the present case in Hyderabad. Nevertheless, the quality of wastewater will be a great concern if urbanisation takes place concurrently with increases in industrial, hospital and commercial effluents. Further, it is noted that most of the wastewater usage in the developing world including India is informal and indirect. This means wastewater is discharged into rivers and the contaminated river water is used for irrigation. Therefore the need today is proper management of the resource, which implies cost-effective and appropriate treatments suited to the end use of wastewater, supplemented by guidelines for their application, as in the case of developed nations, Australia for example. Equally important is farmer and consumer education in risk management strategies as well as improved institutional coordination. Given the situation in India, where self-governed institutions—Water Users Associations—have been established to manage the use of water from the river or canal under the Participatory Irrigation Management programme this is achievable.

Findings from this study clearly indicate that wastewater flow in the Musi River downstream is largely responsible for the better functioning of the WUAs as compared to the WUAs upstream where the river runs dry, except during the monsoons. Therefore, the WUAs formed to manage the canal water can also be made responsible for wastewater management. Further, in India, effective wastewater management necessitates coordination between the urban authorities, water and sanitation agencies, health care agencies, agriculture ministries, urban and industry planning agencies, development and welfare agencies. Therefore, through the WUAs, which are well established and have already developed links with most of these agencies, it will be easy to co-ordinate the activities. This will also ensure participation of the users, which is supposedly a very important aspect of water and wastewater management policies around the world.

However, one of the biggest obstacles in such cases is lack of clarity among the user groups regarding wastewater-irrigated agriculture. The main risks and benefits are not well understood; further, for policy makers, wastewater is not a priority issue. These issues hinder the process of designing an integrated solution. However, with the involvement of WUAs these can be tackled effectively and thus ensure sustainable management and use of urban wastewater for agriculture. Some of the following initiatives should help better management of wastewater in developing countries like India:

- Planning for wastewater source separation and treatment,
- Preventing water pollution by proper management techniques and provision of incentives for wastewater use,
- Developing preventive and curative health care measures, and
- Designing farmer extension services for each category of wastewater dependent group, such as landless labourers, land leasers, landowning farmers, etc.

References

Buechler SJ (2004) A sustainable livelihoods approach for action research on wastewater reuse in agriculture. In: Scott C, Faruqui NI, Raschid L (eds) Wastewater use in irrigated agriculture–confronting the livelihoods and environmental realities, CABI/IWMI/IDRC

Buechler SJ, Devi G (2003) Household food security and wastewater-dependant livelihood activities along the Musi River in Andhra Pradesh, India. Report submitted to the World Health Organisation (WHO), Geneva, Switzerland

Buechler SJ, Devi G, Raschid L (2002) Livelihoods and wastewater irrigated agriculture along the Musi river in Hyderabad city, Andhra Pradesh, India. Urban Agriculture, 14–17 December

Institute of Resource Development and Social Management (2006) Personal communication, Hyderabad, Andhra Pradesh, India. July 2006

International Water Management Institute (2007) IWRM challenges in developing countries: lessons from India and elsewhere. Water Policy Briefing 24

Keremane GB, McKay JM (2006) Self-created rules and conflict management processes: the case of water users associations on Waghad canal in Maharashtra, India. Int J Water Resour Dev 22 (4):543–559

Matriu V (2000) Conflict and natural resource management. FAO, Rome, p 1

McKay JM, Keremane GB (2006) Farmers' perception on self-created water management rules in a pioneer scheme: the Mula irrigation scheme, India. Irrigat Drain Syst 20:205–223

Sivamohan MVK (2001) Pro-poor intervention strategies in irrigated agriculture in India: some issues. In: Hussain I, Biltonen E (eds) Irrigation against rural poverty: an overview of Issues and Pro-poor intervention strategies in irrigated agriculture in Asia. International Water Management Institute (IWMI), Colombo. pp 79–84

Chapter 9
Lessons Learned and Way Forward

Over the past century unprecedented developments like growing population, urbanisation and industrialisation have resulted in shrinking freshwater supplies. The current signs indicate that the situation on the ground is getting worse, and not better. Thus, the water managers, planners and policy makers around the world face the challenge of finding new sources of supply to address perceived new demands. The challenge is more acute because the options for increasing the supplies have become expensive and are often environmentally damaging.

In recent decade as drought and dwindling groundwater supply have become a reality, source substitution increasingly is viable solution to our water supply challenges. It is a suitable alternative to satisfy certain uses, allowing higher quality waters to be used for domestic supply. Accordingly, in many water-scarce regions, water reclamation, recycling and reuse have come to occupy a prominent place in water and wastewater management policies. On the other hand, it is also true that water crisis is a crisis of governance. Discussions in earlier chapters clearly indicate that more efficient participation of formal and informal organizations in the management and development of water is necessary and thus mandatory.

With these thoughts in mind, this research focuses on reuse water in agriculture, policy and implementation in Australia and India. It attempts to grasp and analyse the ongoing multi-faceted problems of wastewater management with a focus on the role of the public sector, the private sector and the community. The research selects three case studies representing different models of governance: the Virginia pipeline scheme (PPP model), Willunga Pipeline scheme (Divesture model); and Musi irrigation scheme (unsupported/informal wastewater reuse). From these studies, it thereby attempts to draw lessons from these experiences by posing the following questions: What are the governance models that aid in implementing sustainable water reuse in formal and informal water economies? Does community social capital contribute to the implementation of a sustainable water reuse project, and in what ways? What follows is the summary of the lessons learned from these case studies.

© The Author(s) 2017
G. Keremane, *Governance of Urban Wastewater Reuse for Agriculture*,
SpringerBriefs in Water Science and Technology,
DOI 10.1007/978-3-319-55056-5_9

The overall objective is to provide guidance to better understand the institutional and governance challenges of managing urban wastewater, particularly for reuse in agriculture. Not only do lessons learned help assess and improve project implementation but they also can help minimize similar issues on similar projects.

9.1   Lessons Learned

The lessons learned are presented in Tables 9.1, 9.2 and 9.3; the aim is to understand the difference in the cases selected, explain environment for wastewater treatment and usage in both countries, and measure social sustainability with respect to the selected cases.

Table 9.1   Key features of the wastewater policy frameworks in Australia and India

Features	Virginia pipeline scheme	Willunga pipeline scheme	Musi irrigation scheme
Type of scheme	Wastewater irrigation scheme	Wastewater irrigation scheme	Surface water irrigation scheme
Regulatory and policy framework for wastewater use	National Guidelines for Water Recycling, 2006 South Australian Reclaimed Water Guidelines, 1999	National Guidelines for Water Recycling, 2006 South Australian Reclaimed Water Guidelines, 1999	WHO Guidelines for the safe use of wastewater, excreta and grey water, 1973 WHO Health guidelines for the use of wastewater in agriculture and aquaculture, 1989
Governance model	BOOT model (public-private partnership)	Divestiture model (private company)	Self-governed institutions
Ownership of the scheme	State (built on the BOOT model)	Willunga Water Company	State owns the Musi irrigation scheme, WUAs responsible for O&M of the scheme
Use of wastewater	Irrigating market gardens	Irrigating vineyards	'Source of livelihood' growing vegetables, rice
Quality of wastewater	Treated to Class A	Treated to Class B	Untreated/partially treated
Beneficiaries	Heterogeneous group in terms of ethnicity	Homogenous group of grape growers	Heterogeneous group (small farmers and landless)
Access to wastewater	Users sign contract with the water company	Users sign contract with the water company	Use in defacto illegal manner

Table 9.2 Regulatory, institutional, technical, financial and socio-cultural environment for wastewater usage in Australia and India

Environment for wastewater treatment and usage	Virginia pipeline scheme	Willunga Basin pipeline scheme	Musi river irrigation scheme
Regulatory and institutional			
Quality standards and regulations	Clear	Clear	Not clear
Freshwater availability at project level	Sever scarcity	Scarce with restrictions	Scarce
Institutional framework	Formal	Formal	Informal
Community, public, and private sector involvement	High level of community, public and private sector involvement	High level of private sector involvement, no public sector involvement	High level of community involvement, no public or private sector involvement
Regulatory and enforcement mechanisms	Strict and strong	Strict and strong	Weak
Technical			
Conveyance and distribution	Sophisticated infrastructure facilities	Sophisticated and innovative infrastructure facilities	Local or primitive conveyance methods
Reliability of reclaimed water supplies	High	High	High
Quality of reclaimed water	Class Á	Class B	Untreated and/or partially treated
Impact on crop yield and use of fertilisers	Positive	Positive	Positive
Health risks	Minimal	Minimal	High
Financial			
Financing the scheme	Pooled effort	Users	No cost
Willingness to pay	Increased WTP	Increased WTP	Not sure
Profitability to farmers	Profitable	Profitable	Profitable

(continued)

Table 9.2   (continued)

Environment for wastewater treatment and usage	Virginia pipeline scheme	Willunga Basin pipeline scheme	Musi river irrigation scheme
Socio-cultural			
Markets for the crops grown using reclaimed water	Well established	Well established	Local markets only
Psychological aversion towards wastewater usage	For direct potable reuse, not for agricultural use	For direct potable reuse, not for irrigation purposes	No psychological barrier for using wastewater for irrigation
Concern for opinion of reference groups and public criticism	Diminishing	Diminishing	Diminishing

Table 9.3   Community social infrastructure in the three study sites

Elements of social capital	Virginia pipeline scheme	Willunga pipeline scheme	Musi irrigation scheme
Symbolic diversity	H	L	H
Resource mobilisation	M+	H	L
Networks	H	M+	M

Note L Low; *M* Medium; *H* High

9.1.1   *Wastewater Policy Framework in Australia and India*

With different systems of governance, the water supply institutions in India and Australia work in two different domains. Likewise, the regulatory and policy framework for wastewater treatment and usage are also entirely different and therefore largely remained incomparable. However, this study attempts to compare the processes of governance and institution formation in these countries. The aim of this section is to compare the wastewater policy framework in Australia and India based on some of the key features presented in Table 9.1. These key features help us to understand how each case differed from one another.

It is clear from Table 9.1 that while the schemes in Australia are identical except for the governance model, the Indian case differs in at least a dozen of issues. First and foremost is the type of scheme: the Virginia and Willunga scheme are exclusively wastewater irrigation schemes, Musi scheme on the other hand is originally a surface water irrigation scheme which over a period has turned into a wastewater irrigation scheme (particularly downstream). This is mainly due to the vast amount of urban and industrial effluents flowing into the river all the year round which will only increase given the current trend of increased urbanisation and industrialisation.

Secondly, the regulatory and policy framework for wastewater use: In Australia, wastewater reuse for non-potable applications is largely formal which means there are rules and policies that are written documents and are executed through formal position, such as authority or ownership. The two schemes studied in Australia had a strong regulatory and policy framework in the form of National Guidelines for Water Recycling, 2006 developed under the National Water Quality Management Strategy, 1992. At the State level, there is the South Australian Reclaimed Water Guidelines, 1999 prepared by the Environment Protection Agency (EPA) and the Department of Human Services (DHS) on behalf of the Environment Protection Authority (the Authority) and the Public and Environmental Health Council (the Council). On the other hand, India has no such clear guidelines and the nation still relies on the WHO guidelines for the safe use of wastewater, excreta and grey water, developed in 1973 and updated in 1989.

Another major difference observed in the cases studied is the ownership of the schemes. The Australian schemes are properly planned reuse schemes while the Musi case is the representation of how wastewater use and management occurs in developing countries like India. The Virginia pipeline is built under the BOOT contract wherein the State retains the ownership of the scheme. The other case in Australia—the Willunga pipeline is owned by a private company. In case of Musi, although the irrigation scheme (canal) is owned by the State/Irrigation Department the Water Users Associations (WUAs) are responsible for operation and mainte- nance of the scheme. Therefore, this makes WUAs responsible for managing the wastewater by default.

When it comes to the priority in use of wastewater, one thing common across all the schemes is that wastewater usage gained importance because of the depleting freshwater resources. A noticeable feature here is that, in Australian cases wastewater appears to be an alternative source to supplement the freshwater resources and the wastewater is treated to a quality matched to particular end uses. In Indian case, large numbers of small farmers and landless depend on wastewater for their livelihoods. This is a cause of concern because all the wastewater that is used In India is either untreated or partially treated, and complete prohibition or adoption of any stringent set of guidelines is not a practicable solution.

Further, the Virginia scheme sells Class A reclaimed water to a heterogeneous group of market gardeners who belong to different ethnicity and are either broad acre or glass house farmers. The Willunga scheme supplies Class B reclaimed water to a homogenous group of grape growers who wanted to catch up with the boom in wine exports. In India, wastewater is either untreated or partially treated, never- theless is a source of livelihood for small farmers and landless labourers.

The users in case of the Australian schemes need to sign a supply contract with the private water company to gain access to the water from the scheme and need to pay a fixed tariff for using the water and services of the company. On the other hand, wastewater use in India is done in a defacto illegal manner and it is free for the users. But, it has associated problems which outweigh the benefits of using it. Apart from these points, the institutional and social environments for wastewater use in both countries are very different which is discussed in the following sections.

9.1.2 Institutional and Social Environments for Wastewater Use

This section discusses the institutional and social environments for wastewater usage in Australia and India. Table 9.2 compares and explains how the observations made in each case reflected institutional and social characteristics of that particular case. Institutional and social features shape the desire and decisions of:

- Users who buy the reclaimed wastewater (in Australia) and use it for irrigation,
- Water provider or water company who sell the water (in Australia), and
- General public who buy the crops watered with reclaimed wastewater.

Having selected schemes with different governance structures, and considering that the study sites have varying socio-economic and political conditions, these findings will help draw lessons that can be useful for implementing wastewater reuse projects elsewhere.

9.1.2.1 Institutional Environment

Implementing a successful wastewater reuse scheme mainly depends on the institutional environment existing in that particular state or country. The regulatory and institutional environment encompasses wastewater quality standards and regulations; regulatory and enforcement mechanisms; institutional framework and involvement of the community, public and private sector.

Wastewater quality standards and regulations—The quality standards and regulations for wastewater reuse for agriculture in Australia are clear while in India they are not. Australia in general and South Australia in particular, where the schemes selected for the study are operating, have clear regulations and guidelines for wastewater reuse. In fact, the favourable regulatory and institutional environment or in other words, strict and consistent regulation has been one of the major reasons for the successes of these reuse schemes. In India, however, the rules and regulations are not in place and we can find unrestricted use of wastewater along the rivers, which receive effluents from the growing cities and industries. Nevertheless, this study finds that if the WUAs along these rivers (as in the present case) are made responsible for managing the wastewater, the problem of regulating wastewater usage can be largely addressed.

Regulatory and enforcement mechanisms—The regulatory and enforcement mechanisms related to wastewater reuse are strict and strong in Australia. This implies that the rules are enforced as formulated and the regulatory authorities are strict when it comes to enforcement of the rules. For example, throughout the study regions we find signboards stating 'reclaimed water being used—not for drinking' and lilac-coloured pipes to indicate the reclaimed water distribution network. The situation in India is the reverse. The regulatory authorities are either weak or in some cases non-existent, and hence when it comes to wastewater reuse regulations and their enforcement, India is way behind Australian standards.

Institutional framework—The countries under study have different water economies. Australia has a formal water economy with public and/or private service providers serving most of the water users. In the case of India, it is informal, and the water users depend largely on self-supply, informal exchanges and local community institutions. Further, the 'rules in use' governing wastewater usage is also well defined and clear for the Australian cases. On the other hand, in India there are 'self-created rules' governing water (canal water) in place. However, for cases like the Musi River where the canal water is nothing but the wastewater discharged into the rivers, either the rules are not stated or there are no rules.

Public and private sector involvement—It is understood that community acceptance and participation is of utmost importance to implementing a wastewater reuse scheme. This implies that due to uncertainties related to water quality issues and negative public perceptions, reclaimed water has not yet found acceptance, particularly for direct potable use. In the present case, it was evident from the two case studies in Australia that community, public and private sector involvement backed up by a favourable regulatory regime can lead to implementation of successful reuse schemes. In both these cases, the involvement of these entities was found to be high. In the case of the Virginia pipeline scheme, it was public-private and community participation; whereas in the case of Willunga pipeline scheme it was total private sector involvement (the community formed itself into a company). In India, it was community organizations represented by the Water Users Associations. Although the WUAs were originally formed to govern canal irrigation systems, in the present case due to the very nature of the Musi irrigation scheme, the WUAs were responsible for managing wastewater use. This was because the river was perennial downstream due to continuous wastewater flows, and hence the WUAs were active, compared to the WUAs upstream where the river ran dry except during the monsoons. So downstream, there was a high level of community participation. However, there was little or no public or private sector involvement. This calls on policy makers to think along the lines of making wastewater use more formal, thereby reducing or minimising the risks associated with the use of untreated wastewater.

9.1.2.2 Technical Environment

The technical environment is also equally important for the success of any reuse scheme. It encompasses the conveyance and distribution system, the reliability of wastewater supplies, the quality of the water, and impacts on health, crop yields. etc.

Conveyance and distribution—Water distribution is a vital component of any irrigation system and this is no different in the case of wastewater irrigation projects. In case of both the schemes in Australia—the Virginia pipeline scheme and the Willunga pipeline scheme, the private companies (WRSV and WBWC respectively) were responsible for laying the pipeline, and for operation and maintenance of the system. These schemes have a modern conveyance infrastructure and water distribution is efficient. Although highly technical, the system is user friendly. In India, however, the wastewater conveyance system is the same primitive system used for canal irrigation using river water. In addition, individual farmers have their own structures and channeling methods for distributing water to their fields, which reduces overall distribution system efficiency because of the poor management practices.

Reliability of reclaimed water supplies—In all three cases, there is high reliability of reclaimed water supplies. Considering that the source of water is wastewater from treatment plants (in Australia) and mixture of treated and untreated sewage from the cities and industries (in India), the supply of wastewater is assured. In India, the very fact that the river remains wet downstream as compared to upstream, where there are no wastewater inflows, indicates that the supply is reliable. Further, in both the countries, with the growth of population and urbanisation the supplies of wastewater will be continuous and increasing in the future. In the Australian cases, there is still scope to increase the capacity of the two schemes and efforts in this direction are under way.

9.1.2.3 Financial Environment

The financial environment relates to the funding of wastewater reuse schemes and also the willingness of the users to pay for this resource, which is regarded as waste. The profitability of using wastewater, as against fresh water, is also an important aspect of the financial environment as it relates to the tariffs of wastewater. All though this study did not focus on the profitability analyses some information related to financing the scheme, and tariffs were collected by reviewing of the project documents (in Australian cases).

Financing the scheme—Acquiring funds to develop a water reuse scheme is an onerous task mainly because of the negative public perceptions about wastewater use. However, through proper planning and well-designed partnerships, these issues can be addressed effectively, as in the case of the two schemes in Australia. In the case of the Willunga scheme, the users contributed the initial funding entirely, with no assistance from the public sector, unlike the Virginia scheme where it was a pooled effort by the SA Water Corporation, Federal Government, Water Reticulation Services Virginia (WRSV), and the Virginia Irrigation Association (VIA). In India, there is no cost component involved in construction of any infrastructures and the use of wastewater is uncontrolled.

Tariff structures—The Australian schemes had a clear tariff structure for using wastewater and in India the farmers paid no fees for using wastewater. However, it is to be noted that generally, the WUAs charge the farmers for using canal water, but not in this case since it was the 'wastewater'. The prevailing water fee structure in case of the Virginia pipeline is seasonal and comprise of Connection fees and Water fees which includes the supply fees and water use charges. Water use charges is billed four times a year—Summer, Winter and the spring and autumn Shoulder Season with varying rates. In the case of the Willunga scheme the tariff structure comprised two components: (1) a fixed tariff based on allocated water entitlement and (2) a variable tariff based on usage. Both the tariffs were cheaper than the mains water supplied by SA Water.

9.1.2.4 Socio-cultural Environment

The socio-cultural environment largely includes the perceptions the public has towards use of reclaimed wastewater to irrigate crops. It includes the markets for the produce irrigated with reclaimed water, psychological aversion towards the use of reclaimed water, and concern for public opinion and the opinion of reference groups.

Markets for crops grown using reclaimed water—Reclaimed water use for agriculture is a widespread practice, and all three schemes under study delivered water for irrigation. As stated before, in India, the use was to grow crops for self-sustenance or for the local market only. The produce marketed was largely paragrass, a fodder grass. There were some who sold vegetables, but the numbers were much less. In Australia, in both cases the market for irrigated crops (using reclaimed water) is well established. The produce from Virginia market gardens has a very good market within the state and a proportion is exported to other states. In the Willunga scheme, the wine produced from grapes grown using reclaimed water has a good market within South Australia and across all states in Australia and abroad. In all three cases the perception of public towards crops irrigated with reclaimed water seems to be positive; and this can be improved even further with proper awareness and education, particularly in India.

Psychological aversion towards wastewater usage—Many of the previous studies on reclaimed wastewater usage in agriculture have studied the human perceptions about wastewater usage. Most of them conclude that there is generally no psychological aversion by the users towards wastewater usage if it is for non-potable uses such as toilet flushing, watering of gardens and lawns, and agri-culture. However, this is only when there are strict and strong regulatory and enforcement mechanisms in place to control and monitor the entire process. The two case studies in Australia agree with these findings, as farmers had no psy-chological barriers towards using wastewater for irrigation. The situation was similar in India; however, it is to be noted that the farmers there are still using untreated wastewater because it is the only source that supports their livelihoods. Having said this, there is a strict 'NO' to using reclaimed wastewater for potable purposes in both Australia and India.

Concern for opinion of reference groups and public criticism—Concern about
the opinion of reference groups and public criticism is one of the disincentives to
users of reclaimed water. However, the cases under study clearly indicate that the
influence of this factor is diminishing in all regions. But in some places these
factors still influence the farmer's attitudes and perceptions towards wastewater
usage. Some of the reference groups that were identified in the Indian case study
were community leaders, religious leaders, and local politicians. So although this is
diminishing it still needs to be taken into account while planning a wastewater reuse
project.

9.2 Wastewater Reuse Schemes and Social Sustainability

Compared to the concept of environmentally sustainable development the concept
of socially sustainable development has received less attention in discussions on
sustainable communities. Generally, studies on wastewater reuse focus more on
environmentally sustainable development which is of course very important. So is
the concept of socially sustainable development. This study focuses on the later and
is more concerned with the development of social capital which is the regulator of
sustainability.

It is true that the initial motivation to seek more sustainable alternatives to
freshwater supplies (urban wastewater in this case) is driven by economic or
environmental or health-related factors. However, it cannot be ignored that
achieving 'sustainability' is not a win/lose event, rather it is a process which
involves constant awareness and ongoing evaluation of the achievement of the
desired goals. Building upon this idea, the findings of this study indicate that a
critical factor linking increased social capital with the implementing of a successful
and sustainable reuse scheme is that the community citizens and irrigators both
begin to see that their (collective) action can make a difference in achieving goals.
Social capital therefore makes a difference in terms of a community's ability to
solve its own problems—the problem of water scarcity in all these three cases. In
addition, it was clear from the case study results that although the governance
structures developed to manage urban wastewater reuse in Australia and India
varied, one thing common in all the three cases was the increase of community
social capital, as measured by level of trust.

Further, based on the concept of entrepreneurial social structure it is apparent
that diverse symbolic structures, wider resource mobilisation, and diverse networks
result in increased social capital which in turn makes a difference in terms of a
community's ability to solve its own problems—water scarcity in this study (see
Table 9.3).

In case of the Virginia pipeline scheme it is noticeable that even with heterogeneous or diverse groups (symbolic diversity) it is possible to achieve increased
social capital and thereby social sustainability. This is further supported by the fact
that the level of trust among the members of the community and different

stakeholders was high. Moreover, after operating successfully for seven years it is still maintained and maintaining social capital means social sustainability. In case of the Willunga scheme, there was wider resource (inputs such as knowledge, time, and money) mobilisation as the scheme is entirely developed and implemented by the users or irrigators.

On the other hand, in India, the scheme studied involves unplanned use of wastewater for agriculture. However, diverse institutions and organizations—Urban Farmers Associations, Water Users Associations, and caste groups (as most of the users were from backward communities) were involved with wastewater use, clearly suggesting higher symbolic diversity. Furthermore, the idea of forming the WUAs was undertaken mostly to ensure high levels of community cooperation and involvement in water management activities.

Communities with entrepreneurial social infrastructure can identify problems and alternative ways to solve them. These case studies, particularly the Virginia pipeline scheme, demonstrate that it is possible to develop and maintain 'entrepreneurial social infrastructure' even with diverse groups. Only then can the communities participate in any change or shift in a positive, proactive way. Combined with an increase in community social capital, this ultimately can be the path towards achieving social sustainability.

9.3   The Way Forward—Policy Options

Today, in most countries around the world, on the supply side of the wastewater market, wastewater collection is well organised and has reached reasonably high levels. Wastewater treatment still needs to be improved, which can be achieved by adopting a demand-driven approach instead of the existing supply-driven approach, allowing for technological innovations, and integrating it with environment and water resources strategies. On the demand side of the market, the regulatory and institutional frameworks are of great relevance in determining the decisions of the farmers who use the reclaimed wastewater to grow crops, and the community who buy crops irrigated with reclaimed wastewater. In addition, technical, economic, and cultural incentives influence wastewater reuse for agricultural purposes.

Consequently, the following recommendations are made to improve the acceptance level of farmers to using the resource and thereby developing a sustainable and successful reclaimed water irrigation scheme. These suggestions are based on the findings from the current case studies in Australia and India:

- Prepare location-specific guidelines for wastewater use and management,
- Ensure private sector involvement and enhanced community participation in wastewater treatment and management,
- Design awareness programmes, on the legal, social, economic, environmental, and health issues related to waste water and target all key stakeholders,

- Design appropriate arrangements and agreements among all those who hold a stake in wastewater management,
- Conduct thorough financial and technical feasibility studies to ensure the scheme's long-term viability and to attract private sector funding,
- Prepare a competent technical design and develop appropriate safety measures and practices to avoid any occupational health and safety hazards,
- Conduct regular water quality monitoring, and control best irrigation practice through soil surveys, regular soil and crop management reports in order to ensure environmental sustainability,
- Decide on a tariff structure that is affordable to the users and also ensures the financial sustainability of the project, and
- Above all, build up a favourable policy and regulatory regime for wastewater reuse based on the local socio-economic and political situations.

From the policy perspective, the following policy options are suggested based on the findings of this study:

Policy option 1: When we consider wastewater markets, the supply side collection and treatment of wastewater are usually under the jurisdiction of a sector (such as urban water supply and sanitation) that is different from the reuse sectors (such as agriculture and municipalities), hence intersectoral coordination in planning and management is extremely important. On the demand side, users should be involved in planning and monitoring the quality of the supplied effluent. Effective advisory/extension services are also extremely important.

Policy option 2: Wastewater use should be viewed with a multi-disciplinary approach so that all parties benefited or affected (public and private sectors, consumers and farmers) can be informed about the benefits and risks of wastewater use, the options available to manage such use more effectively and the livelihood activities of different groups that are sustained by wastewater (in the developing world).

Policy option 3: With all the available modern technologies, it is not a problem to treat wastewater to a quality matched to end uses; however, considering the associated investment and recurring costs that are required to treat wastewater might be a constraint in developing countries like India. In such cases, setting up short-term objective to control wastewater exposure to consumers and producers may be feasible. This can be attained through participatory approaches such as farmer's field schools to educate farmers on crop selection to minimise exposure and safer and sustainable irrigation practices. The benefits of this approach can be enhanced by public health education, therapeutic medical care for irrigators, and community awareness programmes.

Policy option 4: A major policy shift is needed for water management investments that are important for irrigated agriculture. Although the state is the critical driver, civil societies and the private sector are important actors, and can play important roles in promoting treated wastewater reuse. Therefore, it is essential to table a dialogue between all three societal sectors to find workable solutions.

***Policy option* 5**: Effective and sustainable management of wastewater use in agriculture requires developing and applying practical wastewater use guidelines. But, in absence of strict regulatory enforcement to ensure compliance with the guidelines on the part of water authorities, those discharging wastewater, and those handling and using wastewater it is difficult to adopt a set of guidelines developed based on 'no risk' criteria. This is truer in developing countries like India. Therefore, the approach should be to develop and apply realistic guidelines based on 'managed risk' or 'acceptable risk' criteria.